T0246423

The Small Building Contractor and the Client

How to run your business successfully

by Derek Miles

Practical
ACTION
PUBLISHING

An Intermediate Technology Publication

Practical Action Publishing Ltd
25 Albert Street, Rugby, CV21 2SD, Warwickshire, UK
www.practicalactionpublishing.com

© Intermediate Technology Publications 1980

First published 1980

ISBN 10: 0903031671
ISBN 13: 9780903031677
ISBN Library Ebook: 9781780445878
Book DOI: http://dx.doi.org/10.3362/9781780445878

Since 1974, Practical Action Publishing has published and
disseminated books and information in support of international
development work throughout the world. Practical Action
Publishing is a trading name of Practical Action Publishing Ltd
(Company Reg. No. 1159018), the wholly owned publishing
company of Practical Action. Practical Action Publishing trades
only in support of its parent charity objectives and any profits are
covenanted back to Practical Action (Charity Reg. No. 247257,
Group VAT Registration No. 880 9924 76).

Contents

Acknowledgements   4

Preface   5

Chapter 1   Marketing and the Contractor   7

Chapter 2   Estimate or Guesstimate?   37

Chapter 3   Cost and Efficiency   54

Chapter 4   Purchasing   74

Chapter 5   Figuring it out   94

Chapter 6   Putting a Price on it   125

Chapter 7   Bidding Policy   171

Chapter 8   Taking Care   193

Chapter 9   The Builder and the Law   224

Chapter 10   The Contractor and the Client   245

Acknowledgements
The printing of this publication has been made possible by a grant from the Overseas Development Administration. The Intermediate Technology Development Group gratefully acknowledges their generosity.

The illustrations on page 98-102, 105, 108, 151, 158 and 164 are reproduced by kind permission of the International Labour Office. They were first published in *Estimating and Tendering, module 2 of the ILO Regional Course on Construction Management, Nairobi, 1976*. ILO, Geneva. The same source was the basis for some of the material on pages 37, 38, 39, 45 and 46, and permission to use it is gratefully acknowledged.

Preface

Management is about getting things done. Construction management is about getting things built. This book is the result of a decade of experience in helping small contractors in developing countries to get themselves established and to run their businesses effectively. The material has been used as a basis for training courses and should be invaluable for the managers and owners of small contracting businesses interested in improving their managerial capacity. This should in turn help them to become better employers, by offering more permanent jobs, as well as helping their clients by making the contractors more responsive to the needs of their customers.

The problem of 'how to build?', like the problem of 'how to manage?', is as old as man himself. It follows that the management of construction itself has a long history. The story can be traced back to the days of the caveman, whose needs were met by first finding a suitable cave in which to live and then making it habitable through simple modifications. Later, artificial forms of shelter were constructed using local materials such as mud huts with thatched roofs. Then more solid buildings were made using quarried stone, and finally composite and processed building materials, such as burnt bricks and concrete products, were produced to meet specialised demands.

As designs, materials and components have become more complex and demand has risen, the construction industry has taken on a key role and its performance impinges on all other sectors of the national economy. Thus construction costs are not merely a matter of concern to the clients of the industry, but must also concern the nation and its government. Indeed, construction is often responsible for creating more than half a nation's wealth in terms of fixed assets, so value for money must be a prime concern.

Speed of construction, as well as cost, is important. Unfortunately, local construction industries are often criticised by economists and others for failure to meet completion dates and, unfortunately, these criticisms are often justified due to inadequate performance. Indeed there is a vital link between time and money in construction management, and contractors usually find that quicker jobs lead to lower costs and bigger profits.

But governments and international aid agencies have a role in helping their local construction industry to become more competitive. One of the earliest efforts in this field was an Intermediate Technology Development Group project for 'Technological and procedural guidance to the construction industries of less-developed countries'. This was financed for the initial period (1969-72) by the British Ministry of Overseas Development. During this period ITDG co-operated on the development of training material with the Kenya National Construction Corporation Ltd, which was started as a joint venture by the Kenya government and NORAD (the Norwegian Aid Agency) in 1967.

More recently the International Labour Office, with financial support from NORAD, has set up a project to promote the training of practical construction management within the African region. The immediate objective was defined as "to create in the participating countries a basic capability for delivering management training to small scale building contractors", while the longer term objective is to improve the overall managerial and economic performance of the contractors.

The material on which this book is based has been developed over the period. The approach is decidedly practical, with emphasis on providing ideas and techniques which the reader can apply in a straightforward way to increase his knowledge of, and control over, his business. Most of these ideas and techniques are just as relevant to good management in the public sector direct works agency as to running a private business for profit. Saving time — and saving money — are the twin themes.

No book of this kind can come from the innate knowledge of one individual alone. The author willingly acknowledges that he has drawn on ideas and experience from many people and many sources over the years. I would, however, like to mention four people with whom I have had the pleasure of talking and working as co-lecturer, on a number of occasions and in a number of countries. They are Dr Colin Guthrie and Mr John Andrews of the International Labour Organisation and Mr Folkward Vevstad and Mr Jostein Fjellestad, consultants to NORAD. Errors and omissions are, of course, my own. Finally, my thanks to the editorial staff of Intermediate Technology Publications Ltd for their care and effort in preparation.

Derek Miles

Chapter One

Marketing and the Contractor

Who Is the M.I.P.?

Who is the most important person (M.I.P.) as far as your business is concerned? Who is this M.I.P. among all the V.I.P.'s (very important persons) who make a contribution to keeping you in business? Your site staff are certainly important, because they are the people who do the physical work on your projects that results in your contracts (hopefully) being completed on time and within the budgeted costs. Your office staff are also important since they help to plan the business and keep it under control, ensuring that there is a set of definite targets for the site staff to work towards, as well as providing services (such as purchasing, accounting, estimating, etc.) to keep the business going. Then again, your suppliers of plant and materials also make a big contribution to supporting your business activities, as do your sub-contractors. But there is one person without whom you would have no business at all — and that person must be *your* M.I.P. — it is your client.

The Client is a Customer

No business can exist without people who will pay for the goods or services it offers for sale. People will not come knocking on a contractor's door, pleading with him to give them a tender or quotation. The client must be seen as a customer (or, more accurately, *potential* customer) who must be attracted and persuaded that it makes sense to do business with this particular contractor, rather than one of his competitors. Thus the contractor has to be, at least partly, a salesman — and seek out potential clients (and keep existing ones happy) if his business is to continue, let alone expand.

The Business of Building

The starting point for preparing a marketing policy is to recognize the business you are in and who are your potential clients. Building is essentially a manufacturing business, although of a rather specialised kind. Land and a whole variety of different sorts of building materials and components are

fed in at one end of the production line and buildings or roads or bridges come out at the other end. It is a process industry and the builder makes his money by carrying out the process of building efficiently, speedily and at the lowest possible cost. Of course, apart from joinery, precast components and prefabricated buildings, the builder does not operate in fixed factory premises. His 'factory' is the construction site chosen by his clients — and its location inevitably shifts as one contract is complete and another starts.

The Client's Problems

The difficulties that the contractor might meet during the production process are not ones that the client wants to worry about. He will have problems of his own in getting his own work done effectively. What are the client's problems? First he has to get together the land and the finance. Then he will have to recruit professional advisers to decide on the scope of the project and take care of design and supervision. Finally, once the project is handed over, he will have to ensure that the structure is put to good use and that proper maintenance routines and procedures are operated so that it will remain fit for its purpose. With all these problems of his own to cope with, the client will not want to be loaded with additional second-hand problems passed on by his contractor.

The Ideal Contractor

The client will rightly feel that he deserves a rest from problems during the construction period, apart of course from finding the money to meet payments to his contractor on the due dates in accordance with certificates issued and the agreed terms of the contract. From the client's point of view, the ideal contractor is one who will hand over a building of the kind and quality that he needs at the time that he needs it.

The Client's Point of View

The successful contractor gets to the top by trying to see things the way that his client sees them, and always putting his client's interests first. The client does not want or expect to find the contractor on his doorstep complaining that work has come to a standstill because cement is in short supply. It may be true — but it is not the concern of the client. The contractor has signed the contract documents and has therefore undertaken to obtain all the necessary building materials. That is his side of the bargain, and he is legally bound to do it — so it might as well be done without argument or complaint!

Giving a Service

The job of a contractor is to give a service, and in a sense the building business is a service trade like the restaurant trade. If a contractor goes into a restaurant and is not promptly served with a well-cooked dish of food, he will not return (and he is also likely to spread the bad news to his friends). Yet the same contractor may prepare a long list of excuses for poor progress to present at a monthly site meeting to back up a request to extend the contract period, and then be surprised to be dropped from the tender list for the next project. Good clients are hard to find but easy to lose, and bad service or sharp practice are two sure ways to lose good clients.

Looking for Work

Building is a highly competitive business. Clients do not have to look very far to find a contractor who is willing to submit a tender for their latest project. Yet too often the builder seems to think that it is the job of the customer to seek him out and ask for a quotation. A trader in the market place knows very well that without customers he has no business, and he has to out-shout and out-sell his competitors. But if builders are short of work, many either sit at home and wait for something to turn up or write to the newspapers grumbling that times are hard and that the Government ought to do something to help them.

Starting Out

Most small building firms are started as a result of the initiative of one or two brave (or sometimes foolish!) men, sometimes with professional qualifications but often craftsmen, who decide to 'have a go' on their own. Only rarely do such new businessmen stop to think consciously about their 'market', and what sort of clients they want to attract or what sort of work they want to do.

No Strategy

They usually start with little financial backing, and have to grab any work that comes along. Sometimes things work out well. In 1906 a man called Albert Edward Farr erected a sign over his new works at Frinton-on-Sea in England which stated "All kinds of orders executed on the shortest notice". The firm that he started so optimistically is now the successful Bovis Civil Engineering Limited and the subsidiary of a very large group. But today the lack of a clear marketing strategy is a much more serious handicap, as competition in the building business gets steadily more fierce.

'Hand to Mouth'

Even now, very few firms employing up to 50 (or even 100) men really plan their marketing strategy at áll, with the result that they operate on a 'hand to mouth' basis with a wildly fluctuating order book. In turn they cannot retain a steady and loyal work force, keep their plant and equipment fully employed or pursue a long term materials purchasing strategy aimed at securing maximum discounts from suppliers in return for regular orders. The result is that they stand no real chance of breaking through into the big league of construction contractors.

Contract Value
The only limiting factor to most small firms' ambitions is the maximum contract value that they are able to undertake, often stated in the form "We can handle jobs up to $ . . ." Even this vague rule is sometimes broken, and many promising small firms have gone out of business by overreaching themselves and plunging into a major contract without the expertise or financial backing that would be needed to carry it through.

Wild Bidding
The 'contract value' limitation is not enough to save a company from overtrading. Many small firms bid wildly for work of almost any kind up to a certain contract value, and end up with a hotch-potch of different types of work spread over a wide geographical area. The result is that their workload becomes literally unmanageable, mistakes are made both in the office and on the sites and the overstrained managers are too busy to put things right in time. Then losses start to mount up, clients start to complain, the contractor is dropped from tender lists for future contracts — and finally the firm finishes this vicious cycle with a smaller order book and a much worse reputation than it started with.

Self-examination
A preliminary stage to thinking about a realistic marketing strategy is careful and critical self-examination. What do we do well and what do we do badly? What, with extra effort and investment, could we do well? What, barring miracles, will always cause trouble and never yield a realistic return? The answers to these questions will inevitably vary greatly between one firm and another, depending upon the form of organisation, individual expertise and available resources.

What Do We Do?
It is surprising, but true, that many builders have no clear idea of where the majority of their work comes from or what it consists of. They just tender for contracts as they read about them in the newspapers, and do not attempt to specialise and increase their skills in the more promising sectors.

Analysis

The best way to answer the question 'What do we do?' is to take a pencil and paper and analyse the firm's annual turnover for the last year or two, and split it up according to types of project. The result for a typical small contractor might be:

Analysis of Turnover 1979/80

	Work done		Contracts awarded	
	Value ($)	%	Value ($)	%
Housing	108,000	61	26,000	11
Schools	31,000	18	128,000	54
Health Centres	15,000	8	12,000	5
Factories	22,000	12	55,000	23
Roadworks	1,000	1	16,000	7
Total	177,000	100	237,000	100

Significant Information

Even such a basic analysis can yield significant information. It is clear that this particular firm is essentially a building specialist, with only a limited involvement in road construction. It is interesting to note that contracts awarded during the year totalled $237,000 while work done on the various sites amounted to $177,000, which suggests that the contractor's order book is increasing.

Shifting Emphasis

The disaggregated figures suggest that there is a strong shift in emphasis from housing to school construction as the predominant contributor to workload, although factory construction is also becoming more important.

Previous Years' Figures

To obtain a clearer idea of trends it is necessary to compare the latest figures with those for previous years, as in the financial analysis of a company's accounts. If 'work done' and 'contracts awarded' show a steady upward trend over the years, the contractor would appear to be running his marketing strategy on the right lines — providing the firm is reinvesting sufficient of its profits in the business to finance the extra work.

Profitability

But increasing turnover alone is not enough to make a satisfactory business. The contractor's employees would not work for nothing. The contractor's clients also expect some financial return for their efforts. The contractor too will need to ensure that the work that he carries out is profitable, to cover his personal and family expenses and also provide additional funds that can be used for capital investment or to boost working capital and enable the company to take on additional work. So we need to measure profitability as well as workload to obtain a factual basis for our marketing decisions.

Profitability Analysis

If the contractor is running his business efficiently, he should have costs of his operations broken down on a monthly basis for each of his contracts. Since he will also be measuring the work actually completed each month on each site for the purpose of preparing an interim valuation and submitting an interim certificate to the client's representative, the contractor should have a regular monthly check on the profitability of each job. This is vital, since it gives a factual basis for bonus payments to encourage successful site agents, and also gives an opportunity for prompt remedial action when jobs are less profitable than expected.

Example

If these profit figures are available, they can be aggregated by type of contract for comparison with the value of work done over the course of a year. Taking the earlier example, the split of profit contribution from the various classes of contracts might be:

Analysis of Work Done and Profitability 1979/80

	Work done		Profit (Loss)	
	Value ($)	%	Profit ($)	%
Housing	108,000	61	15,000	94
Schools	31,000	18	1,000	6
Health Centres	15,000	8	(3,000)	−19
Factories	22,000	12	4,000	25
Roadworks	1,000	1	(1,000)	−6
Total	177,000	100	16,000	100

Lessons To Be Learned

There are lessons to be learned by the directors of the company even from this very basic set of figures. Housing and factory construction were far and away the most profitable activities for this particular firm, while roadworks and the building of health centres both resulted in losses. School construction was marginally profitable, representing 18 per cent of the work completed during the financial year, but contributing only 6 per cent of the profit earned.

Profits and Losses

It is instructive to look separately at the profit-making contracts and the loss-makers. Housing and factory contracts together made a profit of $19,000 on work done to the value of $130,000. The remaining contracts, involving $47,000 of construction work, resulted in a net loss of $3,000.

Something Wrong?

These figures suggest that there is something wrong somewhere, either in policy-making or in performance. The disparity in returns on the various classes of contract can be shown even more clearly by calculating the profit as a percentage of work done:

Profit as a Percentage of Work Done 1979/80

	Work done ($)	Profit ($)	Return %
Housing	108,000	15,000	13.9
Schools	31,000	1,000	3.2
Health Centres	15,000	(3,000)	−20.0
Factories	22,000	4,000	18.2
Roadworks	1,000	1,000	−100.0
Total	177,000	16,000	9.0

Differing Returns

Overall the contractor is making a financial return of 9 per cent on the work which he carries out. This is a reasonable return, although not particularly exciting. But when we look at the figures more closely, we see that the returns on the various types of contract range from an 18.2 per cent profit on factory construction to a 100 per cent loss on the small amount of roadworks carried out.

Housing and Factories

Both housing and factory construction appear to be well worthwhile, since the work obtained is being completed profitably. In fact it would be possible to drop unit rates to some extent on future tenders of this kind in order to improve the chances of obtaining additional work.

Schools, Health Centres and Roadworks

It is the remaining contract categories of schools, health

centres and roadworks that show unsatisfactory, or even negative returns.

The Reason Why?

Like a watchdog that barks, the figures tell us that something is wrong on these other contracts. But, again like a watchdog, the figures alone cannot tell us the reason why this is so. The firm may just have suffered from bad luck on the health centre work, and the roadworks job may be at such an early stage that the setting up costs have not been recovered from certificated sums due from the client for measured work completed.

Right Jobs? Right Clients?

· But we do know that something has gone wrong somewhere on these contracts. The contractor needs to ask himself two questions. The first is "Did we take on the right work?" The second is "Did we work for the right clients?" If, after answering his own questions as honestly and realistically as he can, he is satisfied that the same troubles could not happen again — all well and good. It will be all right to carry on with the same spread of work.

Time for a Change?

But if either the type of work or the attitude and behaviour of the clients (or their professional representatives) led to the poor performance, it would be time for a change of direction in marketing policy.

Good Money After Bad?

In business everyone makes mistakes sometimes. What marks out the successful businessman is an ability to learn from his mistakes, plus a determination never to make the same mistake twice. There is a great deal of truth in the adage "Never put good money after bad".

Check Past Figures

Once again, the figures for a single year will not be enough on their own as a basis for judging the progress of the company. To get a clear idea of the relative profitability of contracts in the various categories, it is necessary to examine the record of profits and losses on comparable work in previous years. If past figures confirm the validity of those for the current year, we will have more confidence in using them as a basis for policy-making.

The Future

Let us suppose that these past figures did in fact confirm that the rates of return shown in the previous table were reasonably typical. What can that tell us about likely future profitability?

Contracts Awarded

We have no projections for work to be done in the current year, but it is possible to get some idea of probable returns by aplying the percentage profit figures for the various categories of contracts to the value of contracts awarded in each sector. Work may already have started on some of these jobs and some may not be completed by the end of the financial year, but this exercise will at least give us an idea of the likely trend.   ·

Multiplication

The following table gives a simple projection of likely profits (or losses) on current contracts. The figures in the final column are obtained by multiplying the contract value in the first column of figures by the percentage profitability in the next column:

Profit Projection ($)

	Contracts Awarded	Return %	Profit (Loss)
Housing	26,000	13.9	3,600
Schools	128,000	3.2	4,100
Health Centres	12,000	−20.0	(2,400)
Factories	55,000	18.2	10,000
Roadworks	16,000	−100.0	(16,000)
Total	237,000		(700)

Changing Workload

These figures portray an alarming swing from profit to loss as a result of the changing mix of contracts that go to make up the firm's overall order book and therefore its future workload. Projected profits from factory building, schools and housing are more than counterbalanced by the projected losses from roadworks and health centres.

Trends

In fact it is unlikely that the roadworks contracts will do as badly as the projected 100 per cent loss on turnover, and the projections above are based wholly on a continuation of the previous trends. But every experienced contractor knows that losses on one bad contract can swallow up all the profits on several good ones, and the projected figures above show how this could happen. The changing pattern of the order book is shown graphically in the following chart:

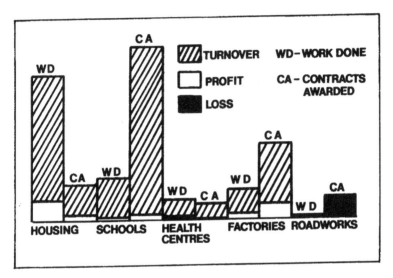

Housing

The chart shows that our example company is losing out on the very profitable housing work that was the mainstay of its order book in the previous year. A well-organised contractor can make good profits on repetitive housing work even if the unit rates are relatively low. Perhaps this firm should consider pricing future housing tenders more keenly as a lower percentage return would be more than compensated by the higher turnover.

Schools

The firm has been very successful in increasing its order book for school construction. However it must hope (and plan!) for a better return than the 3.2 per cent achieved in the previous year. A further point is that over half of the contracts awarded relate to school construction, all of which is presumably carried out on behalf of the Ministry of Education. There is a danger in over-heavy reliance on a single client, since the business is very vulnerable if that single source of new contracts dries up. The strongest firms usually specialise to some extent in a form of technology or a type of contract, but make sure that their workload comes from a variety of separate customers.

Health Centres

Health centre construction was a loss-maker in the previous year. We need to know why. Perhaps it was a single unlucky contract. Perhaps the work consisted of a number of very small buildings scattered around inaccessible villages in remote areas. This is where it is dangerous to base tenders on standard unit rates per cubic metre of concrete, per square metre of blockwork, etc. Transport and supervision costs are always higher for dispersed contracts of this kind, and a single mistake in setting out or a delayed delivery of materials by the supplier can turn an estimated profit into an actual loss. Contracts awarded were to a slightly lower value than work completed, and the proprietor or manager of the firm will have to ask himself whether tenders for this type of work are ever going to produce a positive return. If not, it would be best to leave them for other contractors to cope with.

Factories

This is the most encouraging part of the story. Factory construction, which shows the highest profit margin, also shows an encouraging growth in contracts awarded. Factories are usually fairly simple structures, but they have to be built as speedily as possible since the manufacturer wants to get into production as soon as he can. So a contractor with the management skills to push the work along on the site and guarantee completion within the contract period is likely to find his services in demand. The main danger in working for non-Government clients is the possibility of financial default, so it is vital to make certain that they are credit-worthy, and that the paper profits will actually result in solid increase in the contractor's bank balance.

Roadworks

Naturally the contractor will hope for a much better result than the last 100 per cent loss! But will he make a profit? Building and road construction are two very different activities. They demand different forms of craft, technical and managerial skills. In addition, road construction requires various specialist plant. Successful general builders often sub-contract the road construction on their jobs to specialists in this field, since they would rather forego the possibility of a small slice of the profit than get involved with an activity which they are not really equipped to carry out well. Our contractor will need to ask himself whether he is going to go into road construction in a serious way — since there is no point in diverting his attention into unprofitable sidelines.

Limited Data

The foregoing discussion is based on very limited data on the operations of our sample small building contractor. In real life, we would need much more detailed information.

Other Ways

It might be more informative, for example, to break down the turnover and profit figures in different ways. For example, how do profits on excavation work generally compare with those on concreting? This sort of information is only available if the cost accounting procedures are effective, but it might tell us whether the firm would do better as a specialist sub-contractor concentrating on one particular type of operation than as a general contractor.

Location

Another way of analysing the workload is by looking at the location of the contracts. Do all the profits come from contracts close to the head office and all the losses from remote areas? If so, the estimator should make an appropriate adjustment to rates for transport and supervision on future tenders.

Clients

We could also analyse turnover by client. Are there any clients whose contracts always give rise to losses? If so, in the future rates should be raised or invitations to tender politely declined.

The Result

The result of all this time consuming, often painful, but very necessary self-examination should be a broad definition of the firm's preferred markets in terms of:

1. Type of work.
2. Operating area.
3. Preferred clients.

Type of Work

The preferred type of work will depend on the knowledge and skills of the management and senior staff, but also upon the plant and equipment that is available. Road construction, in particular, demands a range of specialist plant — even if a deliberately labour-intensive approach is adopted. It usually

PREFERRED MARKETS

1. TYPE OF WORK.
2. OPERATING AREA.
3. CLIENT.

pays best to specialise — the 'jack of all trades' really is generally the 'master of none'. Quite apart from improving the competence and competitiveness of the enterprise as a whole, it makes life much easier for the owners and managers!

Operating Area

Many small building firms fail to take full account of the cost of transporting people, plant and materials to remote sites. If the firm has just one truck, it may take a full day to make an urgent delivery to a single site in a distant town while other contracts are held up and have to wait. Small firms tend to operate most efficiently close to their head-quarters, and this is truest of all for the one-man business. If work has to be taken on outside the standard operating area, the profit margin should be higher to cope with hidden costs, loss of flexibility and other risks and uncertainties.

Client

To be realistic and blunt, we need clients who can provide work that can be completed at a profit. No client is going to guarantee a profit to his contractor — except on certain 'cost plus' work where profits are anyway limited. But at least the contractor should be given a chance of profit, providing he performs in a satisfactory way. The experienced contractor will get to know those individuals and organisations that will treat him fairly, be reasonable in their interpretation of contract conditions and make prompt payments in accordance with the requirements of the contract. Good contractors are wise enough to avoid bad clients, just as intelligent clients try to steer clear of bad contractors!

The Marketing Approach

Following upon the process of self-examination and the definition of the strengths and weaknesses of the contractor, it is possible to adopt a scientific marketing approach to obtaining an appropriate workload.

What is Marketing?

Marketing can be simply defined as the method by which a businessman contacts and attracts a circle of suitable customers and clients. He knows that without a market outlet for the product or service which he has available, he has no business at all! There is no point in hiring skilled men, buying expensive machinery and equipment or learning the specialised skills of business planning, organisation and financial management if there are no contracts to manage at the end of the day.

Profitable Clients

But clients alone are not enough. What the contractor (or any other businessman) really needs is 'profitable clients'. The worst sort of 'unprofitable client' is the client who fails to pay. But clients can also be unprofitable if they keep changing their minds about the design of the building, and yet refuse to meet reasonable claims for variation orders or extra payment.

Know your Clients!

The art of marketing is to know your clients, where and how they operate and whether they are both willing and able to make payments to honour interim and final certificates in accordance with the terms and conditions of the contract. In short, the contractor needs to know that his potential client is 'a good person to do business with'.

Know your Competitors!

The builder needs to know his competitors as well as his potential clients. What are their strengths and weaknesses? Are there are gaps in the market which they are not equipped to fill? Is there any method of building in which he can offer a special service that the other contractors in his area cannot match?

Package Deal?

Would it pay, for example, to link up with a local architect or engineer to offer a full package deal or 'design and construct' service? A private client who needs a new building quickly might be prepared to negotiate a package deal contract with such a firm, thereby by-passing formal tendering procedures and giving the contractor a better chance of profitable work.

The Unique Selling Proposition

Advertising men sometimes talk of the u.s.p. — the unique selling proposition. Can the contractor offer some form of unique service to potential clients that no one else can offer? This can take the form of special technical or managerial skills or simply prompt service and flexibility.

Prompt Service

If a local supermarket was burned down, the owner would be desperate to get back into business before his customers got too used to shopping elsewhere. It would be well worthwhile for him to pay over the odds for quick completion. Thus a contract to rebuild might be negotiated at a generous price if a contractor could be found who could work at high speed (perhaps even day and night) to get it reconstructed and back in business.

Success Breeds Success

What is more, if such an 'emergency contract' is completed successfully the client may well decide to award all future work to the same contractor on a negotiated basis at a fair rate of profit. As in life generally, proven success is the provider of opportunities for future success.

Building Up Goodwill

The time for a contractor to start thinking about his reputation is when he starts out in business. Good reputations cannot be bought off the shelf. They can only be built up slowly over the years, by a history of honest trading and fair dealing in small as well as large matters. A reputable building contractor is one who is known and mentioned in casual conversation by his existing clients, as a fair businessman who employs conscientious workers. The aim of every builder should be to gradually obtain a good name and reputation and thereby build up that vital, if intangible, asset that goes under the heading "goodwill".

Advertising

Some builders think that marketing is just another word for advertising. Although large construction firms can easily afford to advertise in newspapers and technical magazines in order to become better known, this is usually too expensive for the smaller ones. In any event, most clients rely on the advice of their professional advisers — architects and engineers — who will make recommendations based on their own judgement, supplemented by factual prequalification data and the candid "off the record" opinions of those of their colleagues who have had direct experience of working with the contractor before.

The Best Advertisement

Selling one's service to new clients can be both tough and expensive. Keeping existing clients is also tough — it means completing jobs on time and to price. Clients can be difficult. But it is vital to keep the client happy, since every building contractor can — and must — afford the best advertisement there is — a satisfied client.

A Satisfied Client

Past and present clients talk, and they talk to potential clients of the future. What do they say about you as a builder?

Are they happy with the service you provide?

Have you a reputation for starting and finishing projects on time?

Do they commend your standards of quality and finish?

Do your clients end with a building which merely satisfies the conditions of contract or are they really pleased with it?

Above all, do they believe that you possess the most important asset for any businessman — a reputation for integrity?

See that you have a good name with anyone and everyone who can give you business in the future.

Satisfying the Client — From the Start

Sometimes the contractor deals directly with the client. On other occasions he is required to work with the client's representative — usually an architect or engineer. Either way the contractor would be wise to establish from the start the client's requirements and priorities and do his best to meet them. The contractor who does this *in addition to meeting the terms and conditions of the contract* will thereby be offering his client an all-important "little extra" which will improve relations between them.

Integrity

Where a client requires technical advice, it may be necessary for the contractor to help him to translate his requirements into what is practicable and give a general indication of what the possible alternatives are likely to cost. In offering this service the contractor must offer honest advice. If he tries to fool his client by putting forward inflated costs, he will eventually be found out and his reputation for integrity and fair dealing will be lost for ever.

Giving a Service

These are some of the ways in which a contractor can offer a service to his client:

1. Establish clearly the objectives and key requirements for the project.
2. Examine the site, and give a rough estimate of costs for alternative layouts.
3. Explain exactly what will be involved in carrying out the work, access requirements and (if the project is for an extension to an existing building) any disruption that will be caused during the contract period.
4. Forecast starting and finishing dates in accordance with existing workload.
5. Give a firm quotation in accordance with contract documents.
6. Be punctual in attending site meetings and inform the client (or his representative) promptly of discrepancies in the drawings or any extra works that seem necessary or desirable.
7. Submit interim and final accounts clearly and promptly.
8. Ensure that the client (and your employees) understand that instructions should come only through official channels, with signed extra works orders where alterations are required.

9. Keep the site tidy and do not borrow tools or equipment from the client without permission.

10. When the job is complete, inspect it with the client and ask if he is satisfied.

Presenting Your Firm's Personality

Good relations with clients come first. But there are other ways of presenting your firm's "personality" in a favourable light. One such is headed notepaper. It is not very impressive to receive a letter from a fellow businessman typed or even hand-written on an untidy scrap of paper. If letters are well-typed on good quality paper with a printed and attractive letterhead, it helps to give the impression of an efficient and orderly business.

Printed Letterheads

Everyone notices a good letterhead, but architects do more than most because design is their business. To make a good impression it should be well-designed (try asking a friendly architect for advice!) and easy to read. Although printing costs may seem high, contractors should remember that a thousand sheets of headed paper will last a considerable time and the cost per sheet is quite low. Good headed notepaper also (and sometimes falsely) gives an impression of financial stability, and can help when seeking a credit account from suppliers or an overdraft from the bank. If only one extra job or other business advantage is gained, all the printing costs will be paid for.

Name Boards

Most builders like to put up a name board at their sites, if only to ensure that materials are delivered to the right place. If the name board is neatly painted and clear it will also serve as a useful and cheap advertisement. It pays to put up a name board even on quite small jobs as it will be read or noticed by the people who pass by, and some of them will remember your name when they next want some work done. Of course the reverse also applies, and a bad sign will give the opposite impression to that intended.

Plant and Vehicles

The same applies to plant and vehicles. If a lorry is rusty, dirty and falling apart, the builder would be well-advised to hide his responsibility by painting his name in the smallest letters possible. Too often we see such lorries proudly displaying the name of the builder, and the public gains the reasonable impression that he will be quite happy to produce inferior and careless work.

CLEAN AND WELL-MAINTAINED PLANT AND VEHICLES ARE AN EXCELLENT ADVERTISEMENT FOR THE FIRM.

Plant to be Proud of

We seldom see a successful firm with shoddy plant. Successful builders know that the cost of a few tins of paint and a few hours of labourers' time spent in washing equipment down once a week is cheap advertising as well as good maintenance. One tip — if you have plant to be proud of and want to paint a slogan on it — is to keep the message short, since it will usually have to be taken in at a glance while the vehicle is moving. A good place to put the contractor's name, in bold letters, is on the tailboard of a lorry or pick-up.

Picking a Symbol

One aid to easy recognition is the adoption of a symbol or special form of lettering to represent the firm. Once it is chosen, use it consistently. See that it appears on your notepaper, printed forms, signboards and the sides of vehicles and plant so that it will become really well-known.

SEE THAT YOUR SYMBOL APPEARS....

ON NOTE-PAPER,

SIGNBOARDS

AND THE SIDES OF VEHICLES.

Using the Press

Paid advertising is expensive, and will only be effective if it is carefully thought out and placed where it is likely to be read by a reasonable number of potential clients. Probably the most effective way a building contractor can advertise is to join with a commercial client in sponsoring an advertising feature when a building contract is complete. This is read as it is regarded as "newsworthy" by the reader, so it helps both the client and the contractor — but it particularly helps the contractor by linking him closer to his client.

Free Advertising

The shrewd contractor can sometimes manage to obtain free press advertising by keeping in contact with industrial correspondents. If you are carrying out an unusual job or one which involves exceptional craftsmanship or will be of great value to the community, it may well be newsworthy. In return for the time taken to visit the newspaper office, you might easily obtain publicity that will bring your name before hundreds of potential customers.

Seeking Out Work

An ambitious contractor who is anxious to expand his business rapidly will certainly have to decide on some form of procedure for seeking out new clients. How this is to be done will depend to some extent on whether the preferred clients are in the private or the public sector. Public sector organisations, such as Ministries of Works or local authority architects' departments, are usually closely circumscribed in their bidding procedures. Private commercial or manufacturing enterprises often give greater freedom to their buying departments, and here it is necessary to build up a good reputation with the decision-making individuals concerned if you are to be invited to tender.

The Client's Representative

Building and civil engineering are both highly technical fields, so the non-technical client will normally rely heavily on his architectural or engineering consultant when drawing up tender lists and awarding contracts. Thus the would-be established contractor must become accepted in the eyes of the local architectural and engineering community as an honest and competent performer who will not let them down if recommended for work. If a contractor fails to produce good work on time, he will damage the professional adviser's reputation (as a result of making a bad recommendation) as well as his own.

A Compact List of Clients

The building contractor has one major advantage over other businessmen. His market is compact. Each individual contract that he undertakes (leaving aside minor jobbing and repair work) is usually worth thousands of dollars (or the equivalent in local currency). Thus he can build up quite a reasonable order book with a fairly compact list of clients.

Pinpointing Potential Clients

The manufacturer of say, toothpaste, is not so fortunate. Since each tube of toothpaste sells at less than a dollar in the shops (and brings him in only a few cents as it leaves the factory gates), he has to reach and persuade many thousands of individuals to give him their custom if he is to make a living. So while the toothpaste manufacturer will seek to reach tens of thousands of potential customers by spending lavishly on advertisements on TV or in the press and magazines, the builder has the different, and less costly, problem of finding a way of pinpointing and reaching the handful of influential people who are in a position to provide him with profitable work.

Individual Approach

Clearly he will need to adopt some form of individual approach. This is likely to mean that the principal of the firm will have to take the time to visit really promising contacts at their offices, although a preliminary approach could be to send letters containing information about the contractor's experience and capacity to a wider circle of potential clients as an introduction.

Who to Approach?

Marketing men tend to talk about 'campaigns', and it is

true that the ambitious contractor has to fight every inch of the way to enlarge his list of clients. So his marketing campaign has to be as carefully planned as that of a general who is fighting a war. In this case, of course, his potential clients are not 'the enemy' — we hope they will end up on the same side as the contractor! But the market itself has to be attacked, and to do this successfully we need to know the people concerned:

— Who are they?
— Where are they?
— How many of them are there?
— Why do they want buildings?
— What will they use them for?

Making Contact

There are many ways of building up a target audience for your marketing campaign. The first step is to write down all the people you know personally who are in a position to put contracts out to contractors. Then ask your staff to add their own contacts — besides possibly adding useful names it will make them feel more involved in building the business. Then try your relatives and friends. After that one can add to the list by looking through the telephone directory or any local industrial directories, together with government establishment lists which will give the names of people to contact in ministries and parastatal bodies.

Be Methodical!

Be methodical about it. Get the names and addresses exactly right. You want something from the people that you are going to approach — as yet they don't necessarily want anything from you. So show them respect by making sure that their names, job titles and addresses are correct.

Don't Forget the Secretary

If you are in doubt about names, responsibilities and job titles in a large organisation — telephone and check with the *secretary* of the person concerned. Speak to her pleasantly — she will be her boss's 'guardian angel' and will almost certainly be in a good position to secure an interview for you with your prospective contact.

The List

Make sure that the list of contacts is typed out and numbered so that no one is missed out and no one is approached twice — which itself would give an impression of inefficiency. Include two additional blank columns on the right hand side of the paper. In the first you will note the date that an introductory letter is sent. In the second you will note the date that the individual concerned is first met in person. After that stage, separate files will be opened to cover all correspondence with each of the most promising of your new potential clients.

The Introductory Letter

You are trying to put over the impression (we hope correct!) that your firm is efficient and will cope competently with any contracts that are awarded. As in a court of law, you will be judged according to the evidence you put forward. The first item of evidence that your prospective client will see is your introductory letter — and that will be the basis for his decision on whether to take your approach further by making a positive response. So the letter must be clear, well-written and attractively set out on good quality paper.

A Standard Letter

It is worth working out a standard introductory letter, which can serve as a model to be varied slightly to suit your assessment of the client's individual requirements. A sample standard letter for a small building contractor is set out on the following page, and can be used as a guide in setting out your message in a logical way. Sentences are generally short, because these are easier to read.

Make It Stick

Remember your letter will be only one of many that finds its way onto your reader's desk. You have to make sure it attracts him and sticks in his mind.

Points to Remember

Particular points to remember (numbered in the sample letter for ease of reference) are:

1. Don't forget the *date* — it shows you are a contractor who is conscious of time and so should get work done on time;
2. *Name and Job Title* — get them exactly right; like everyone else your reader has an ego — so show you care about him;
3. *Subject* — Either 'AVAILABILITY FOR BUILDING WORK' or better still the name of a specific job that the prospective client will have available for tender shortly.

Ace Contractors Ltd.

181/182 Longfellow Road
Shortman
AQUA 9391
Anytown

Telephone 964300 8431.

1. Date
2. Name
   Job Title
   Address

   Dear Sir,
3. Re: (SUBJECT)
4. Confirming my telephone conversation with . . . . (IF APPLICABLE), we would be most grateful for the opportunity to tender for the above work (OR 'building contracts that you may commission' IF A GENERAL ENQUIRY).
5. We aim to serve our clients by offering tenders and quotations that are both competitive and realistic. Our prices are keen because we are a small firm with low overheads, employ a highly skilled work force and enjoy excellent relations with our regular material and plant suppliers. Our experience can be judged from the attached list of projects for which we were the main contractor. All were completed within the stipulated contract period.
6. We are real contractors, not 'paper contractors'. Key operations such as earthworks, concreting and carpentry are carried out by our own full-time staff. Sub-contractors are used only for specialist tasks, such as electricial and mechanical work.
7. There is no need to take our word for it. Please talk to our clients. We recently completed a (e.g. warehouse) in your area for . . . . , so we are well aware of local ground conditions and other problems. Mr — of that company will speak frankly to you about his experience with our organisation if you telephone him at xxxxxxx.
8. We are builders and we stick to what we know. But, in association with local professional architects and engineers, we can also offer a 'design and construct' service. This speeds things up, and ensures that our experience is available to you right from the moment that the building starts to take shape on your drawing board.
9. A letter can only tell a little about our capacity and determination to serve our clients to the limits of our ability. Perhaps you would like me to call at your office and tell you more. Our motto is 'we seek to serve'. We sincerely hope you will give us a chance to use our building skills to serve you.

   Yours faithfully,   ENCL: LIST OF PROJECTS

The subject of the letter should tell the reader *in a few short words* what to expect in the body of the letter that follows;

4. The first paragraph is introductory. It is intended to get the reader into the right mood to absorb your message, and also tell him what to expect in the rest of the letter. It helps, as in the first part of the sentence, to link the letter to any previous conversation or contact that might have been made. The second part tells the prospective client in a straightforward way that you are a contractor looking for work. There is nothing worse than a letter which leads the reader to start with the question 'but what does he *want*?' in his mind.

5. This is the start of the 'message'. We make the most of our good points and turn being 'a small firm' into an advantage by suggesting that we will therefore provide a better service. The list of past projects (which is discussed later in this chapter) should be provided on a separate sheet of paper. This list is important as evidence of our competence, but should not be incorporated in the letter as it would detract from a clear exposition of our message. Naturally any claim such as 'all contracts were completed within the stipulated contract period' should only be made if it is true!

6. Clients prefer contractors who have the resources to offer a direct service, rather than 'wholesalers' who merely pass on the work to others and take a cut of the profit. This paragraph shows that you recognise the client's legitimate concern, and infers that you realise that quality matters as well as cost.

7. This introduces a personal note. It shows that you understand local problems, and you are not afraid (and we hope you are not!) of prospective clients hearing about the experiences of past clients. Naturally you should first obtain the permission of your previous client to act as referee but, providing relations with him were good, he will usually be prepared (and may even be flattered) to be asked to act in this way. (It also provides a legitimate reason to keep in contact with him — in case there is a possibility of follow-up contracts!)

8. This offers a 'little extra' in comparison with the services available from other builders. A package deal 'design and construct' service can speed up the provision of a new building considerably. This could be attractive to a manufacturer who urgently needs a factory extension to expand

production of a profitable product or a supermarket operator who is anxious to start earning profits from a new store opening. For the contractor, the obvious advantage is that he can negotiate reasonable prices and unit rates rather than tender in the dark against an unknown number of competitors. Needless to say, this paragraph would not be included in a letter to a client's professional adviser, who would see the prospect of losing his design work to a competitor!

9. The final paragraph restates the message succinctly and suggests follow-up action. The letter will probably be read just once by the prospective client, so the message conveyed must be the simple one that here is a competent contractor with a sufficiently professional approach to serve his client effectively and well.

Vary to Suit

The sample letter is included as a general guide. Naturally the contractor will vary the contents to suit the circumstances of his own firm, and his assessment of the needs of his prospective clients.

List of Projects

The letter mentioned a list of projects successfully carried out by the contractor. The prospective client will want to know what the contractor has actually built, and who are the clients he has worked for. Naturally the contractor will obtain permission before mentioning them in this way. The list will detail the client, the title, address and value of the project, the contract period together with an amplified description of the work where this is helpful:

ACE CONTRACTORS LTD.
MAJOR CONTRACTS COMPLETED IN 1979

Anytown Housing Project   for Anytown Development
Value: $120,000   Corporation
                                  Contract Period: 10 months
(15 houses with associated road and sewer works)

Warehouse   for Grocery Distributors Ltd.
Value: $83,000   Contract Period: 8 months
(Reinforced concrete frame building with corrugated asbestos roof)

Extension to Anytown Secondary School
Value: $59,000   for Ministry of Education
                                  Contract Period: 9 months

Factory Extension	for Top Engineering Co. Ltd.
Value: $48,000	Contract Period: 5 months
Rural Health Centre	for Ministry of Health
Value: $24,000	Contract Period: 5 months
Hotel Renovation	for Sunset Bay Hotel
Value: $24,000	Contract Period: 3 months

Follow-up

If there is no response to the first approach, don't give up. At best, it is doubtful if more than one in ten of such approaches lead to an immediate invitation to tender. The main thing is that you have made an initial contact, and that can be referred to next time you write to or meet the person concerned.

Leave a Card

Never miss a chance to make new contacts on a personal basis. Always carry some neatly printed visiting cards giving your own name, together with the business name and address of your firm. You never know when you might get into conversation with a possible client or a practising architect or engineer. It is usually possible to mention your firm casually in the course of conversation, and leave them with a card as a reminder.

Keep Trying

If at first you don't succeed, try, try, try again. Too many small contractors twist this good advice to read 'If at first you don't succeed — give up!' A contractor has to be a bit of a salesman, and might do worse than take his lead from the archetypal insurance salesman — 'there's no man with more endurance than the man who sells insurance'. Keep trying. Be persistent with potential clients — but in a pleasant way. Sooner or later they will give you a chance. When they do, grab it with both hands — and show them what an excellent practical building service they have been missing all this time!

Planned Marketing

Remember that there is a limit to the amount of work that any contractor can both take on and execute satisfactorily. That limit is set by the human, physical and financial resources that are available within the firm or could be mobilised at very short notice. Most successful contractors aim for growth in their businesses — but that growth needs to be steady and continuously matched by expansion in staff, plant and the availability of working capital.

The Budget

Planned marketing can only be based on some form of realistic budgeting process. Budgets for sales and production are discussed in more detail in the companion volume *Financial Planning for the Small Building Contractor.*[1] The important thing is to quantify your sales budget, so that you know how large your order book should ideally be from month to month during the coming year in order to achieve optimum performance.

Optimum Performance

The phrase optimum performance is an important one. A contractor can go bankrupt through having too much work just as easily as he can through having too little. With too much work, or 'overtrading' to use the accountant's term, the management is overstretched, plant is not available when it is needed, clients complain about slow progress and — quite possibly — the bank account is overdrawn and the flow of money to pay wages and suppliers' accounts dries up. Even if the firm survives, it will have made much less profit than if it had been less greedy and kept within a workload that it was properly equipped to carry out.

A Target for Orders

For a building contractor, the size of his order book depends mainly on the results of competitive bidding — and the contractor has no way of controlling the bids made by his competitors or the ultimate choice of the client. Thus it is

1. Published by Intermediate Technology Publications, 9 King Street, London WC2E 8HN, U.K.

never possible to keep the order book precisely in line with the hopes expressed in the sales budget. But it should be possible to use it as a target for orders — and consequently as a guide when tendering.

A Flexible Approach

It is foolish to be too rigid and always expect the same profit margin on every job. Often a more flexible approach pays dividends. If actual orders (i.e. contracts awarded) drop too far below the budget, then it might well pay to lower the profit margins incorporated in tenders for a while, so that turnover grows sufficiently to at least recover expenditure on fixed overheads. On the other hand, if the contractor finds that more tenders are awarded than expected, margins on future bids can be increased or invitations to tender for work that is likely to prove difficult to execute may even be politely declined. Most clients would prefer a contractor to state honestly that he is fully committed, rather than take on work and allow it to overrun because his resources are overstrained.

Chapter Two

Estimate or Guesstimate ?

Buying and Selling

Buying and selling is one of the oldest practices in the world. For many centuries in almost every country merchants have brought their goods to the market to sell for money, or alternatively to barter for other merchandise. Always there are two parties to the transaction. One sells, the other buys. Supply and demand are the basic elements in business.

Price

For the transaction to take place, there must be agreement between the parties on what each is to give and each is to take. In a barter transaction, it will cover the quantity and quality of goods to be exchanged. In a cash transaction, the agreement will cover the quantity and quality of goods or services to be provided by the seller, and also the price he will require to be paid by the buyer to make the sale worth his while.

Competition

If two or more fishermen land their catches on the same dock and offer them for sale to the people of the village, they immediately encounter a new problem — competition! If one fisherman was on his own, he would be able to charge as high a price as he liked and make a big enough profit to leave his boat and stay at home for a few days. The only limit to his decision on pricing would be the level at which he knew that nobody would be prepared to buy fish. He would be in the same position as a specialist building contractor in a small remote town when a small contract in his own speciality comes up for bidding. He knows that his only real competitors are large contractors in the capital city who would be faced with very heavy overheads for transport and setting up costs if they put in a bid. Since there is no real competition, the client will have to accept the local contractor's tender unless it is ridiculously high or else abandon the project.

Building is Competitive

Few building or civil engineering contractors enjoy such a monopoly position. Building is a fiercely competitive industry in most countries, with the result that builders head the league tables for bankruptcies and company liquidations rather than high profitability. Profit margins are often low and a small mistake on pricing a tender document can make all the difference between profit and loss. Thus the skills of estimating and tendering need to be understood by anyone who hopes to make a career in the contracting industry.

A Simple Tender

To see the principle behind a simple tendering decision, let us go back to the fishermen who have just landed their catches on the quayside. They have to decide on the price at which their fish are to be sold. To do this to the best advantage they have to guess not only what is the maximum price that their potential customers would be prepared to pay, but also what price their competitors are likely to ask.

The Customer is King

In a competitive market the customer is 'King'. If the man at the next stall offers the same quality of fish at a cheaper price, he will attract the customers and get his fish sold. This is logical. Everyone tries to buy at the cheapest possible price, providing the goods are of acceptable quality. Thus customers (or clients) usually consider a range of offered prices carefully before coming to a decision to purchase. A tender is in fact an offered price and, when the fishermen shout out the prices at which they sell, they are putting forward simple tenders to potential clients.

Other Tenders

All businessmen are involved in tenders of one sort or another. All clients and customers are involved in evaluating them. Let us suppose we visit a new town, and stroll through the streets looking for somewhere to eat. Outside the door of each restaurant is a menu, showing which dishes are being served and at what prices. Again a series of tenders bidding for our custom as potential clients. Each restaurant owner or manager has had to estimate how much he should charge in order to cover direct costs, overheads and profits, bearing in mind the alternative offers being put foward by his competitors.

A Department Store

If we walk into a department store, we are presented with a whole series of tenders for a wide variety of goods. The manager has had to work out acceptable prices for all the goods on offer. If his pricing is unscientific, the customers will quickly seek out the unprofitable (for him) bargains (for them) and leave the remaining high priced items on the shelf.

Not Just for Building

The above examples show that the process of estimating and tendering is not exclusively related to building and civil engineering construction. It is a necessity for the businessman in every walk of life. Without some reasonably scientific way of setting a price structure for the supply of goods and services that he is in a position to provide, no businessman can survive.

The Difference

The crucial difference between estimating and tendering for building contracts and setting prices in shops and restaurants lies in the nature of the transaction. Providing a shop-owner or restaurateur has a good costing system, he will soon know if particular items are showing a loss and will be able to raise his prices immediately to a more suitable level (subject to Government price controls in some countries). Equally, if a certain range of goods is not selling due to undercutting by competitors, it is a simple matter to write out some new, lower price tags or offer a special discount to improve turnover.

Lack of Flexibility

The building contractor does not enjoy this degree of flexibility in setting prices. Once he has signed his name to a contract, he has to carry it out in accordance with the terms and conditions laid down. It has to be completed within the agreed time limit and he will receive no more than the contract price (subject to variations and fluctuations if applicable), whether it is profitable or not.

The Penalties for Failure

But consistently tendering too high means that the contractor will quickly run out of work. If he is short of work, with men and plant standing idle, and finds that his bid for a new project has been rejected in favour of a lower offer, it will be too late to go back to the client with the offer of a special price reduction or discount. So tenders must be neither too low nor too high; they must be just right. Since flexibility is limited, the penalties for inaccurate estimating are high. One serious pricing mistake can soon be put right by a shopkeeper, but could bankrupt a contractor.

The Work of the Estimator

The scientific approach to tendering is to base the tender on an *estimate* of costs. Once costs have been accurately estimated, the estimate can be converted into a tender figure by making an appropriate addition to cover overheads, risk and profit. Skill in estimating is a prerequisite to success for a building contractor. Good estimating does not of itself guarantee success, because good site management based on technical knowledge and craft skills will also be needed to make the contract pay. Contracting is quite difficult enough if the estimator has done his job well and allowed for all reasonable costs and contingencies. But if he has not taken care and the estimate is too low, no miracles of site management can save the firm. A good estimator at least means that survival and success is a possibility. Thus skill at estimating and tendering is a key to prosperity in the building business.

Neglect of Estimating

It is surprising that many contractors give little attention to the problems of estimating correctly. Too often they will leave this to a private individual who "specialises" in producing estimates for builders. Some of these are of course qualified quantity surveyors and others who do an honest job. But other so-called "estimators" may be quite unscrupulous or incompetent.

Guesswork

Contractors need to know the difference between an estimate and a "guesstimate". Too often contractors base their tenders not on an estimate, but upon guesswork. This can be very costly in terms of losses. The contractor should beware of making wild guesses when preparing his estimate. They are expensive in the short term and in the long term will ruin the contractor.

Contractors as Gamblers

Too often contractors seem to have the gambling instinct. They have in their heads a magic figure of a set cost per square foot or square metre. For them estimating at its most scientific means calculating the area of the proposed building and multiplying by their own special magic figure. This figure

is (in their eyes) so sacred that it can be used for building anything anywhere — from a mud hut next to their home to a multi-storey hotel on the top of a mountain. They are, in commercial competence, as bad as the gambler who picks a horse with a pin and is disappointed to see it finish last in the race.

Minimising Risk

Risk will always be a part of contracting. It is a high risk business. But the shrewd contractor foresees risk, allows for it and constantly seeks to minimise it. The professional gambler minimises his risk by studying the qualities and form of each horse in the race, the peculiarities of the racecourse and the weather conditions. In the same way (but with a more worthwhile purpose) the professional contractor minimises his risk by basing his estimate on detailed and accurate information and an analysis of past experience.

Forecasting the Future

The estimator has the job of forecasting the future. Before submitting his tender the contractor has to decide *now* on what it will cost to carry out work in six or twelve months' time. Naturally the future for an industry or a company, as for an individual, is uncertain. We cannot be absolutely certain where we will be or what we will be doing tomorrow or next week or next month. But we can *estimate* all these things on the basis of past experience. What is certain is that the results of a more scientific, information-based approach to estimating produces far better results than wild guesses. Practice makes for proficiency, if not perfection, and experienced estimators can get their forecasts uncannily close to the eventual, actual measured costs.

The Objective

Before starting on any activity it is wise to set down on paper exactly what one is trying to do. The objective of using a system of scientific estimating as a basis for the preparation of tenders is to obtain jobs on which the contractor has a chance of showing a reasonable profit. It is *not* just to obtain jobs at any price. Work that is almost certain to show a loss is better avoided. Naturally no estimator can positively guarantee that his cost forecasts will prove accurate, so that the hoped-for profit margin will be achieved. Actual achievement will depend on good cost-conscious site management, but a realistic set of unit cost estimates will provide a target for performance and a reasonable basis for comparing the performance of agents and foremen on different sites.

Quantities and Costs

Constructing a building is not easy. It involves a large number of difficult decisions about which materials and components to buy, interpretation of drawings and specifications and the management of labour and plant. If you wrote down all this in words, it would be necessary to write a long book to describe even the simplest building. The job of the estimator is not to put the building process down in words. It is his job to get it down in numbers. He needs to reduce the whole process of building to figures representing *quantities and costs.*

The Difference between Estimates and Tenders

The contractor must remember that there is a fundamental difference between an estimate and a tender. The estimate represents the best forecast that the contractor can make of exactly how much it will *cost* him to construct the project for which he is to submit a tender. If he managed to get it exactly right and submitted a tender at the estimated cost, at the end of the contract he would have made neither a profit nor a loss.

Tendering for Profit

But a contractor is rarely content just to avoid making a loss. He hopes to make a profit — not just to provide some personal spending money, but to provide finance to expand

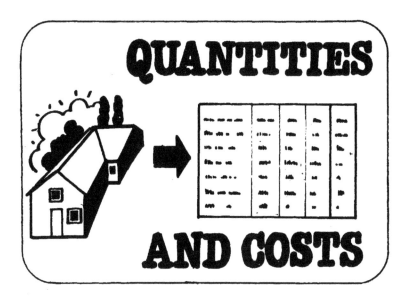

and re-equip his business and cover the many risks that are involved in contracting. Thus the completion of the tender form is a completely separate stage which follows the preparation of an estimate. It is a management job to review the estimate and make a decision on tender prices based on:

— Expected workload.
— Site management availability.
— Availability and cost of finance.
— Overheads.
— Required profit in relation to risk.

Three Steps

In fact the process of obtaining work for a contractor (as distinct from carrying it out at a profit) can be divided into three separate stages:

1. Obtaining an invitation to tender.
2. Preparing the estimate.
3. Review the estimate and complete the tender.

Step Two

Every one of these three steps is important, but it is often the second step — estimating — that is the most daunting and worrying for the small contractor. The thought of reducing a building into numbers and areas and volumes and weights of various kinds of materials and components just gives him a headache and he doesn't know where to start. In fact "taking-off" quantities is not so difficult providing you go about it

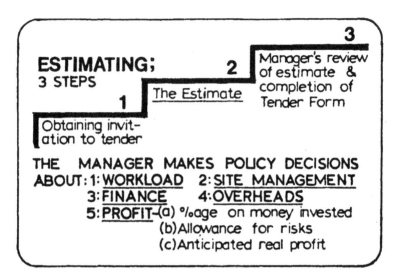

ESTIMATING;
3 STEPS

1 Obtaining invitation to tender

2 The Estimate

3 Manager's review of estimate & completion of Tender Form

THE MANAGER MAKES POLICY DECISIONS
ABOUT: 1: WORKLOAD   2: SITE MANAGEMENT
3: FINANCE   4: OVERHEADS
5: PROFIT (a) %age on money invested
(b) Allowance for risks
(c) Anticipated real profit

methodically, and it has to be done if guesstimates are to be replaced by estimates. No one can price a non-standard building just by looking at the drawings. But it is much easier to put a price on say, 40 cubic metres of concrete, 3,000 blocks, 5 flush doors, 6 light fittings, etc. Sometimes these *quantities* are provided for the contractor as a part of the tender documents in the form of a *bill of quantities*. But the contractor is usually expected to check that these figures are correct by doing his own independent taking-off from the drawings and documents.

Unit Prices

Whether the quantities are provided or not, the contractor will get no help from the client in deciding unit prices and applying them to the quantities to calculate the total sum to be submitted as a bid to the client. The experienced contractor will draw on a combination of factors — including cost records from similar jobs, quotations and other market information, productivity data, specialist advice, etc.

Estimating Productivity

Labour wages and salaries, together with associated on-costs, are major contributors to contract costs. Good levels of productivity and output must be achieved consistently if these costs are to be recovered. Yet many contractors have little clear idea of what levels of productivity they are achieving or need to attain. You do not need a stopwatch and a diploma in work study to get an idea of productivity levels; all that is needed is a little intelligent observation.

Learning Observation

Soon after one of my friends started on his career in construction, an old carpenter took him aside and said: "Son, if you want to go anywhere in this business learn the cost of doing the business. Learn to observe what is taking place around you". Then he pointed to three bricklayers working on a wall. He said "Remember where they are now. Then come back after an hour has passed and work out how much they have achieved". Knowing the area laid and the wall thickness, it was simple to calculate the number of bricks laid. Division by three gave the typical output of a bricklayer in that firm. Bringing in the hourly wage and oncosts for a bricklayer (plus the labourers who assist him) yields a typical cost of laying bricks per hundred.

Simple Calculation

The calculations are so simple that advanced mathematics and a pocket calculator are not required. Observation is the key both to estimating and to the effective control of site operations. Of course, one observation alone is not enough. A careful contractor will regularly keep an eye on the main indicators of productivity, such as the cost per cubic metre of excavation, the cost per square metre of formwork or the time taken for a carpenter to hang a door. By making a regular check on such indicators, he will also know whether the general levels of efficiency in his firm are advancing or declining.

What Makes a Skilled Estimator?

What are the qualities and qualifications that make a man a skilled and reliable estimator? Understanding arithmetic and the ability to make simple mathematical calculations quickly and accurately is a good start, but these qualities alone are not enough. An experienced artisan will know a good deal about basic costs associated with the practice of his own craft, say carpentry, but may be less reliable when it comes to estimating costs for excavation or concreting. Three crucial qualifications must be a wide experience in the area of the construction industry in which the contractor operates, a feeling for costs and sound judgement (sometimes called "common sense" — but nevertheless a quality often in short supply!). In addition the estimator should look ahead by watching and judging likely future movements in prices and wages, and should have a good practical understanding of how "money works". In short, in addition to basic qualifications, he must have a flair for his task.

Where do you find him?

Naturally estimators with all these abilities and qualifications are hard to find — which perhaps explains why so many contractors base their tenders on "guesstimates". To start with, the contractor ought to know something about costs himself. Even if he is unable to prepare an estimate and tender without help, he should at least be able to judge whether the prices that are being put forward are reasonable. A contractor entering the building business without a good grasp of basic costs stands about the same chance of survival as an unarmed man walking into the lions' den. But if the contractor does not have the skills — or the time — to prepare his own estimates, he will have to hire one. The cheapest estimator will not necessarily be the best; if he cannot value his own skills properly, he is hardly likely to put a proper value on his employer's skills when doing his job! Competence and honesty are worth paying for, and may not be easy to find. Let us suppose we decide to advertise for an estimator — what qualities and qualifications should he be required to demonstrate?

The Ideal Estimator

The following list of qualifications would ensure a thoroughly competent — if not an ideal — estimator for a general building and civil engineering contractor:

1. Broad practical experience within the construction industry. Familiarity with the overall process of construction, together with the ability to break it down into component parts and define and price every element of cost on a specific project.
2. A basic grasp of engineering theory, with a quantity surveying or cost engineering qualification and practical estimating experience.
3. Site experience of the practical management of construction contracts, with particular emphasis on cost measurement and control.
4. Analytical ability to evaluate alternative methods of construction and resulting costs in evaluating a job using drawings, specifications and notes from a site inspection. A grasp of contract law in order to understand the implications of the conditions of contract.
5. Capacity to carry out a large number of detailed calculations quickly and accurately in order to produce a thorough and comprehensive cost estimate.

6. Sufficient general experience of planning, programming, resource scheduling and costing to reliably forecast manpower, plant, materials and working capital requirements.

7. Experience in building up a 'library of actual unit costs' from completed projects as a framework for comparison when deciding on estimated costs.

8. A record of scrupulous honesty in business dealings (so that there is no danger of the contractor's tender prices being divulged to, or influenced by, competing firms).

Getting Value from Your Estimator

Now that we are employing an estimator, how should we use him? Since he is an expensive man, it is tempting to keep him busy by overloading him with a large number of tenders to prepare. This is not really a sensible policy. The preparation of an estimate requires a painstaking and careful approach and something is likely to be forgotten if the process is rushed. It is particularly important that the conditions of contract are read thoroughly, and time is given for discussion of difficulties that might be met in carrying out the work within the contract period. None of these decisions should be rushed. Remember that one contract taken at too low a tender price can lead to disastrous losses and ruin a contractor. So give your estimator time to do his job properly, and don't overload him.

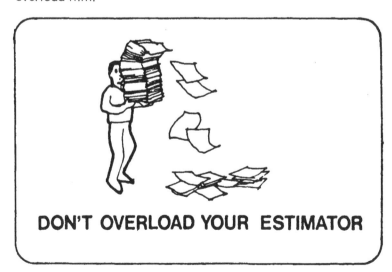

DON'T OVERLOAD YOUR ESTIMATOR

The Three Steps to Tendering

Producing a tender is a mixture of factual forecasting and management decision. In many small businesses the general manager also prepares his own estimates, simply because there is no one else within the firm upon whom he can rely for this vital task. But even when the manager and the estimator are in fact the same person, it is important that he separates out the technical and the managerial tasks in his own mind. So that we can see clearly how the "estimator" and the "manager" share the task of obtaining work for their firm, let us return to the "three steps to tendering":

1. Obtaining an invitation to tender.
2. Preparing the estimate.
3. Review the estimate and complete the tender

Obtaining the invitation to tender

Before the estimator can start work, it is necessary to secure an invitation to tender from a potential client or his architect or engineer. Sometimes this is just a matter of scanning newspaper advertisements. But it will always be vital at some stage before the contract is awarded to satisfy the client that the contractor is competent and able to carry out the work as required within the time available. Just as the contractor would prefer to deal with a builders' merchant who has a good name for delivering the right product at the right time, so the wise contractor will give a good deal of thought to ways in which he can build up a good name in the business community. This is mostly a question of marketing, which was discussed in Chapter One.

A Job for the Manager

Marketing policy must be decided by top management. Getting a good name for reliable work depends largely on the performance of site agents, foremen and craftsmen. But in seeking out new opportunities to tender, the manager must also take into account the capacity of the firm to carry out additional work. Has it sufficient finance, managerial capacity, skilled labour and plant?

Get the Documents in Good Time

It is also important that the tender documents should be obtained in good time, so that the estimator can do a thorough job. If the documents are delivered at the last minute and the preparation is rushed, it is all too easy to quote inaccurate prices or make foolish omissions or arithmetical mistakes. So the contractor should always use the maximum time for submission allowed by the client.

Preparing the Estimate

Once the documents are received at the contractor's office,

they are passed to the estimator. His job is the factual one of *estimating building costs*. His skill is to see that these forecast costs will still be accurate and can be worked to in several months' time, if and when his firm is awarded the contract and has to carry out the work. It may be that he will call in the site agent or general foreman who would be responsible for supervision to advise on possible techniques and methods even at this early stage.

Completing the Tender

The final stage is to review the estimate and complete the tender. This is essentially a managerial decision, involving the evaluation and balancing of risks and acceptable levels of profit. The manager will have to take into account how badly the contractor needs the work when deciding on the minimum acceptable profit. Of course the lower the profit element that is added to the basic unit rates, the lower the tender and the better the chance of being awarded the contract. It may sometimes be necessary to accept very low profit levels just to keep the firm going and cover overheads, but in the long run reasonable profits are required to cope with unforeseen problems and provide funds for expansion. The actual level of profit that can be commanded as a percentage of turnover depends mainly on what the contractor has to offer in terms of skill, efficiency and reputation — another reason for paying attention to marketing.

Incentives and the Estimator

Most staff respond to financial incentives — but make sure the incentives are working in the right direction! For example, it would be wrong to offer the estimator a bonus on the basis of the work he may "obtain" for your firm by preparing "successful" estimates. This would encourage him to quote unrealistic on-site prices (and blame site management when they didn't work out in practice). The job of the estimator is just to present a *factually-based* forecast; it is for the manager to interpret them and make a final decision.

Planning and Estimating

Some contractors think that the time to start thinking about planning the job is after the contract has been safely awarded. They can see no point in wasting time on producing a plan at the tendering stage for a contract that they may never get. Of course it is better to start planning at the award stage than not to plan at all, and too many contractors still run their jobs on the basis of trial and error — with plenty of trials and plenty of errors!

Planning as a Way of Life

The successful contractor has a more positive approach. For him (or her), planning is a way of life — and the most satisfactory way of coping with the inevitable risk and uncertainty that every contractor faces. Since the risk and uncertainty of a contractor's life starts at the tendering stage, he argues that contract planning should start there too.

The Basis for the Estimate

The pre-contract plan need not be as detailed as the working programme, but the estimator will only be able to do his work thoroughly if he has in front of him a plan of some kind which shows the how, why and when of what will have to be done physically on the site. He needs to be fully satisfied about methods, resources and project times in order to prepare his priced bills of quantities and quote a realistic completion time.

Planning and the Client

The intelligent client will also be interested in seeing that his prospective contractors have based their tenders on plans rather than hopes. A contractor who has no plan to cope with risk and uncertainty is very likely to run into trouble — and trouble for the contractor usually means trouble of some kind for the client, even if the contractor alone has to bear the consequent financial loss.

A Draft Programme

Thus more and more clients are requiring that bidders submit a draft programme for carrying out the works as an integral part of their tender. In evaluating the bids, the client (or his professional adviser) will examine the priced items in the bill of quantities together with the stage completions set out in the programme (usually in the form of a bar chart) to ensure that the contractor has really thought about the job and come up with a proposal that can be followed with a reasonable chance of success.

Time is Money

For the contractor, even more than for most other business-men, time is money. The bill of quantities (together with the unit rates and resource breakdowns upon which it is based) provide a cost budget for the job. The bar chart (or critical path programme) provides a time budget, and therefore target times for controlling the completion of each component part of the job. By making regular weekly and monthly checks to ensure that the job keeps within both cost and time targets, the site supervisor should be able to keep the whole job on target.

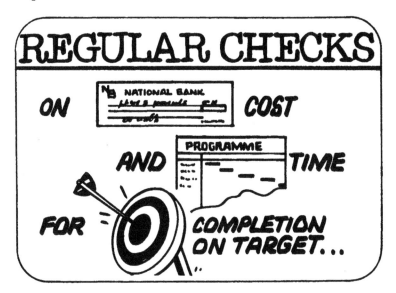

How to Plan

Every contractor worthy of the name should be able to produce *and use* a simple bar chart to control his projects. A simple explanation of programming techniques is contained in the companion volume *Financial Planning for the Small Building Contractor*.

Estimates for Control Purposes

A lot of work goes (or should go) into producing an estimate. Yet many contractors forget about the estimate once they have sent off their tender, and will start from scratch if the contract is awarded on deciding how to plan and control the work on the site. To save wasted time and effort, the estimate should be prepared in such a way that the information it contains can be readily used for contract planning and cost control. It is worth giving a good deal of thought to devising a simple costing system that will provide a uniform basis for accounting/costing procedures throughout the business.

Prompt Costing Essential

The precise system will vary from business to business,

but it is essential that the information should be as up-to-date as possible — even at the cost of being less than 100 per cent accurate. You need to know that a budget head is likely to be overspent in time to put things right on the site — an absolutely accurate "post mortem" examination after the job is over is no use in guiding site management.

A Simple System

The costing and accounting system needs to be simple enough for the contractor himself to understand. If it is not possible to take home any file or record book at night, read it through, understand it and spot any mistakes — then the system needs simplifying.

Overheads

The treatment of overheads and oncosts is a tricky problem, and the method chosen must be consistent from contract to contract. Labour oncosts can be recovered by a straightforward percentage addition to wages payments. Plant and transport costs are best recovered by a system of "internal hire charges" in which sites are charged for using equipment on an hourly, daily or weekly basis — and the plant department operates as an independent profit centre. Central office charges will probably have to be recovered from individual contracts by a percentage charge proportional to turnover.

Work in Progress

Building contracts take a long time to complete with the result that, at any given moment, a large proportion of working capital is represented by "work in progress". If work in progress is overvalued, the contractor will imagine that his contracts are more profitable than they really are and may therefore put in tender prices that are too low when bidding for future work. Work in progress can be valued either on the basis of cost (including overheads) or by measurement using unit prices as set out in the tender. The safest method of valuation is to use whichever of these alternatives produces the *lower* valuation. Providing that the contract is profitable, this means basing the valuation of work in progress on a calculation of costs plus relevant overheads.

Calculating Profit

If the above method is adopted, credit for profit earned will only be taken at the end of a job. Except where contract periods are very long, this should mean that the accounts are realistic.

Provisions

The only exception is where there is a good reason to suppose that the later stages of a job are going to lose money or the contract will overrun, giving rise to liquidated damages. Let us suppose a $500,000 contract is half finished and has so far, at $250,000 just barely covered its costs. In this case the estimator would be required to estimate the *cost* of completing the work plus any liquidated damages that might be incurred. Suppose that he works out that at current prices the remaining work will cost $350,000 to complete and that there is likely to be an overrun of four weeks — leading to a charge of $20,000 for liquidated damages. The contractor has to finish the job in accordance with his contract — and now he knows that it is likely to shown an overall loss of $100,000 plus $20,000 or $120,000. Thus he should immediately make a *provision* of $120,000 in his accounts to allow for this. Such an item sometimes comes under the heading of "contingencies".

Chapter Three
Cost and Efficiency

A Definition
One definition of a contractor is 'someone who can do for 50 cents what any fool can do for a dollar'. This is not a completely fair definition, because building is becoming more technical year by year and a contractor consequently needs a stock of specialist knowledge, experience and skills that 'any fool' would be unlikely to possess. But the definition still has its merits — in that the mark of the professional contractor is his ability to carry out work in his specialist field at a lower cost than his less capable competitors.

The Crucial Link
There is a crucial link between cost and efficiency. The client wants an efficient contractor who can produce a building to the specified quality standards within an accep-table contract period. These two basic performance criteria are seen by the client as elementary, and would be expected of every contractor who is invited to submit a tender. The client normally makes his final choice between this group of generally qualified contractors on grounds of price alone. Every intelligent client naturally wants to pay as little as possible for the facilities which he is commissioning, just as every intelligent contractor buys his materials from the cheapest supplier who offers an acceptable service.

Price and Cost
For the contractor, estimated cost plus profit equals his bid — or the price the client will pay if he accepts the bid. A contractor who is less efficient and so can only build at high cost will end up by only obtaining work when he cuts his profit margin to the bone or even works at a loss. No one owes the contractor a living — like every other businessman he has to sell his services in a competitive market place.

Unit Rates Vary
Contractors are living very dangerously if they take the short cut through the tendering process by taking on work at 'standard rates' extracted from some handbook or passed

round by word of mouth. The unit rates that matter to Contractor X are the rates at which *his firm* can carry out work — and they may be completely different from a list of realistic rates for Contractor Y.

Specialisation

An earthworks specialist may be able to quote very competitive rates for cut and fill on a roadworks job and still make a profit on that part of the work. But his steelfixing, formwork and concreting rates on the bridges and culverts will have to be higher than average since he lacks experience and will need to recruit specialist staff and purchase additional plant and equipment. He may even have to sub-contract this part of the job, in which case his bid will be based on the most competitive quotation received from qualified sub-contractors. The important thing is that his estimate should reflect the most accurate forecast possible of his own costs in carrying out the project, and that his bid should allow a reasonable profit margin on top of this.

Performance Targets

If accepted, the bid will represent his *price* to the client for executing the work, and his **estimated costs** will provide performance targets for his site staff at each stage of the contract. If they can keep costs within the estimates, the hoped-for profit should emerge. If they can, by increased efficiency, cut actual costs below the estimated figures, the final profit figure will be even better. Either way, providing that the estimate is realistic, the eventual profit will be an accurate measure of the contractor's operational efficiency.

Past Experience

If we are presented with a complex set of drawings for a power station, a secondary school or a new motorway, only a genius or a fool would attempt to put an accurate price on it straightaway. All that we could reasonably do would be to "guesstimate" a general figure, to an accuracy of say plus or minus 20 per cent, on the basis of past experience of similar projects. This sort of disciplined guess at an overall project cost is not nearly good enough for a contractor, who has to be much surer of estimated costs before he puts his name — and his bank balance — at risk with a firm bid to the client to carry out the work.

Order of Magnitude

This sort of general forecast of the 'order of magnitude' of the project cost can be very helpful to both the client and his consultant at the early briefing stage of the project management process. At this early stage, before the detailed design has started, a general idea of the likely project cost is sufficient for decision-making purposes in evaluating alternative courses of action.

For the Contractor

A rough 'order of magnitude' cost forecast is certainly not sufficient for a contractor to commit himself to a contract price, but it may be helpful as a check on more detailed estimates based on a breakdown of the overall project into elements. If the final result of the detailed estimates agrees reasonably closely with the 'order of magnitude' forecast, all is likely to be well. But if it is distinctly lower, the detailed breakdown should be checked to make sure that no part of the job has been forgotten. It is sometimes difficult 'to see the wood for the trees' in producing a detailed estimate, and a double check before sending in the bid may save the contractor from being committed to a bid which will prove unprofitable. A good carpenter 'measures twice and cuts once'. A good estimator 'checks twice and bids once'.

Component Costs

The essential feature of the science of estimating is the ability to break down the overall project into manageable and costable units. The overall project is not 'costable' in the sense that one can make a reasonable guess at the total cost of all the various material, labour, plant and other resources that will be required in order to construct it. But as we break the project down into components, it becomes progressively easier to put a price on the separate parts. It is easier to price a single classroom than the whole school. It is easier to price one wall than the whole classroom. It is easier still to put a realistic price on a single concrete block, a cubic metre of mortar and an hour of a mason's time than to get an immediate price for the whole wall.

An Accurate Forecast

By putting accurate unit prices on each of these component resources and then putting all the components together, we end up with a reasonably accurate forecast of the cost of constructing the whole building.

The Bill of Quantities

The basic document used in estimating construction costs

is the Bill of Quantities, which is usually prepared by the client's professional adviser and issued to the competing contractors on major projects, but which might have to be prepared by the contractor himself if required for his own purposes on smaller jobs. On large projects there are likely to be several separate bills, each covering a separate aspect of the work such as foundations, walls, roofing, finishes, etc as well as one called "preliminaries" and covering setting up and general costs such as insurance. Each bill contains a large number of individual items such as:

"Supply and fix 20 gauge galvanized corrugated mild steel roofing sheets as specified."

How Much? and How Many?

The purpose of the Bill of Quantities is to tell us 'how much?' or 'how many?' we need of each item or component that goes to make up the whole building. Even when it is supplied to the contractor as part of the tender documents, he is normally required to check it for accuracy and completeness as errors or omissions are specifically excluded as subjects for claims. The contractor "prices" the bill by putting a unit rate against each item, then multiplying the unit rate by the stated quantity to produce a price for the item. The total price for each bill is brought forward to a summary sheet where a grand total for the job is calculated, usually with a specified percentage addition for "contingencies" which will cover extras and other unforeseen costs to be incurred on the authority of the client's representative.

The Estimating Process

Estimating can be seen as a process. The first stage is to break down the complete project into broad categories, each of which can be represented by a separate bill. This breakdown is often most conveniently done by trades, together with an initial bill to cover preliminary and general items. Separation by trades is particularly helpful when subcontractors are involved. For a typical building job, the bills might be:

1. Preliminaries.
2. Groundworks.
3. Concreting.
4. Walling.
5. Roofing.
6. Carpentry.
7. Ironmongery.

8. Electrical Work.
9. Painting and Glazing.

Once the bill headings have been decided upon, the next step is to list the items in each bill. The number of items into which the work covered by each bill is to be separated is a matter for judgement, and will depend mainly on the likely overall cost and the complexity of the job. The next stage is to "take off" the quantities for each item from the drawings. The unit of measurement will be chosen according to the nature of the item. Sewer pipes will be measured according to their length in metres, while the excavation and backfill will probably be measured in cubic metres and there may be separate items for excavation at various depths. There may also be contingency "extra over" items for excavation in soft rock and hard rock, which would be paid to the contractor only if this was actually encountered.

Description, Unit and Quantity

At this stage in the process, with all items listed together with descriptions, units and quantities, the contract goes out to tender and the would-be contractors start to get involved. It is for the contractor's estimator to calculate unit prices for carrying out the work involved in each item, based on his best forecast of the cost of the necessary materials, labour, plant and equipment as well as overheads. He will then, using the simple arithmetic of multiplication and addition, complete the bills of quantities and the summary sheet. A decision will also have to be taken on the required margin of profit, and whether this should be taken as a uniform percentage addition to the estimated unit cost on all items or "loaded" more heavily on some items than on others.

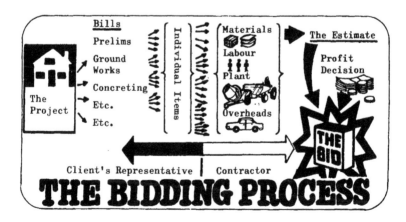

THE BIDDING PROCESS

Comparative Estimates

A short cut to providing unit prices can be the technique of comparative estimating, in which one simply applies unit prices that proved accurate on previous jobs to a similar item on the contract that is open for bidding. It can be a useful technique, and save a lot of time in pricing items such as the placing of concrete or the laying of blocks, where costs and productivity do not vary too much from one job to the next. But it can be dangerous if applied unthinkingly, because no two jobs are really exactly the same. For example, soil conditions vary considerably from one site to the next, materials costs rise the further one gets from the source of supply and skillful, hard-working casual employees are sometimes very hard to recruit locally.

Pre-contract Estimates

Comparative estimates can be usefully employed to give the architect or engineer a good approximation of the project cost prior to sending the project out for tender, and this information will also help the client in making preliminary arrangements for detailed financing at the pre-contract stage. But contractors should be very careful in applying the technique of comparative estimating, since they have to "put their money where their mouth is" and any error could prove very expensive.

Cost Feedback

For the contractor, comparative estimates must be based on actual costs resulting from site cost returns. Too often contractors simply apply unit rates which they have used in previous tenders without ever bothering to find out whether they are realistic. Ideally the estimator should try to build up a "library of unit rates" from site cost returns so that he has a feedback on the accuracy of past estimates as well as an indication of the likely range within which future unit costs are likely to fall. By regularly comparing estimated unit costs with actual unit costs, the estimator will get a better idea of the key factors that affect costs on various operations. As these regular cycles of estimate — actual — estimate — actual progress, so should the gap between estimated and actual figures on each job diminish.

Analytical Estimating

Although comparative estimates provide a short cut to unit rates, the only sure way to an accurate estimate is by a full analysis of the work and resources that go into each item.

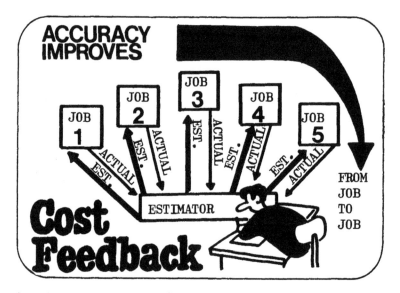

ACCURACY IMPROVES

JOB 1 · JOB 2 · JOB 3 · JOB 4 · JOB 5

EST. · ACTUAL · EST. · ACTUAL · EST. · ACTUAL · EST. · ACTUAL · EST. · ACTUAL

ESTIMATOR

FROM JOB TO JOB

Cost Feedback

It takes time and careful work, but it also provides the contractor with a much better insight into the costs of running his business.

A Typical Item

As a simple introduction to analytical estimating, we can examine a typical item in the bill of quantities for a building — the construction of a concrete block wall. This might appear as an entry in the bill of quantities as follows:

Description	Unit	Quantity	Rate	Total
Walls				
190 mm thick concrete blocks bedded in mortar as specified	M^2	220		

Faced with this item, the contractor needs to decide on the costs involved per square metre of concrete block wall before multiplying by 220 to obtain the cost of building the 220 sq. metres of concrete block wall called for on the drawings.

Cost Elements

To start with, let us list the expenditures that are likely to be involved in a typical item. Then we can go on to put figures against each subhead of expenditure:

1. Labour cost — Wages, allowances, national

	insurance, holiday and sick pay schemes and other direct oncosts;
2. Materials cost	— Cost of materials and components including transport to the site, together with allowance for wastage and other losses;
3. Plant cost	— Plant, equipment and tools relevant to the item concerned;
4. Overheads	— Supervisory costs, rents, services, insurance premiums, car expenses, petty cash, etc;
5. Profit	— Usually calculated as a percentage return on the value of work done by the contractor.

Direct and Indirect Costs

Labour, material and plant costs can be generally classified as *direct costs*, since they are mostly directly related to the work to be done and rise or fall as the volume of work rises or falls. This is not of course precisely correct unless the contractor depends completely on casual labour and hired plant, but it can be taken as a reasonable approximation. Overheads and profit are *indirect costs*, in that they are not directly related to the work done on the site. Overheads can be further split into site overheads which cover the foreman's salary and site office, etc (some of which may be specifically covered in the 'Preliminaries' bill items) and general overheads, such as head office costs and management salaries.

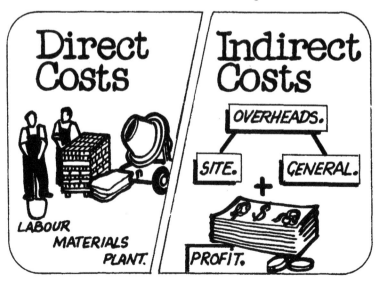

Rate for Concrete Block Wall

As a simple introduction to the procedure for calculating unit rates, let us decide on the cost elements that go into the rate per square metre for constructing a 190 mm thick concrete block wall as required in the example item above.

Labour

We need to know how much we will have to pay in terms of wages and direct oncosts for a mason (with labourers to assist him) to build one square metre of wall. In calculating unit rates, we have to work out the cost of constructing just one square metre of wall but do so on the assumption that this one square metre is only a small proportion of the complete job. If we were in fact pricing a very small job where only a single square metre of block wall was to be built, the rate would in fact have to be much higher to cover setting up costs, etc. In the present example, a total of 220 square metres of wall are required so our unit cost per square metre will reflect the fact that the mason will have a reasonable amount of work ahead of him and so will be working at a good tempo.

Working Rate

The basis of the labour estimate is a forecast of the number of labourers who will work with the mason, and how much they will get through — between them — in the course of a single working day. It is best to start by estimating the working rate per day, because one can allow for meal breaks and other interruptions that are inevitable on a building site more easily on a daily than an hourly basis. At this stage cost and time records from past jobs are particularly valuable as a guideline to what might be achieved in the future.

Forecast Rate

Let us suppose that we forecast — on the basis of records of past levels of output plus an assessment of the working conditions on the new job — that one mason assisted by two labourers can build 8 square metres of wall per day.

Wages and Oncosts

We now know the number of people to be employed on building the wall and how much of it they are likely to build in a day. To work out how much this activity will cost we naturally need to know how much the labour itself will cost on a daily basis. Remember that labour cost is not necessarily the same thing as wages. In many countries there are labour

EXAMPLE

1 MASON

+

2 LABOURERS

COMPLETE 8 SQ METRES PER DAY.

laws that require compulsory contributions for national insurance and sometimes there are taxes directly related to payroll. These must be added to basic wage rates since the cost to the contractor can only be recovered from work done on the site.

Holidays and Sick Pay

If paid holidays are allowed and wages are continued when workmen fall sick, these costs must also be recovered. If an employee is allowed two weeks paid holiday per year, it means that he gets paid for 52 weeks but only works for 50 weeks. Thus we should multiply his basic wage oncosts by $(\frac{52}{50})$ so that we recover sufficient money in the 50 weeks that he does work to cover the costs in the other 2 weeks when he is being paid but produces no return. The same principles applies to sick leave, although here we have fore-cast a likely figure which will apply to the whole labour force, since it will naturally vary from individual to individual. If, with a labour force averaging 50, a total of 250 days or 50 weeks paid sick leave is incurred each year, then we can recover the cost by assuming that 1 week of paid sick leave will be incurred by each employee each year. This means that the actual productive working year for the average employee is only 49 weeks and the "holidays" multiplier above should be modified from $(\frac{52}{50})$ to $(\frac{52}{49})$ to recover the 3 weeks work that will be lost each year by the average employee. Al-together this represents a little over 6 per cent of wages and oncosts in the example.

Overtime and Bonus Payments

If premium rates (1¼, 1½ or even 2 times basic rate) are paid for overtime on contract work, the rate used in estimating must be adjusted to suit. Bonus payments must also be allowed for, although where bonus payments are strictly related to increased productivity, their cost should be covered by the increased output that will be achieved. In other words, a bonus payment of 25 per cent to the mason and labourers in our example will be no problem if they can increase their working rate from 8 sq. metres per day to 10 sq. metres per day.

Example Wages and Oncosts

For the sake of the example, let us assume that the mason receives $6 per day in direct wages and additional charges amount to 45 cents. We assume that the two labourers each receive $3 per day and additional charges amount to 30 cents. Thus their daily cost to the firm will be as follows:

1 Mason @ $6	6.00	
additional charges	.45	
	6.45	
holiday/sick pay @ 6%	.38	6.83
2 labourers @ $3	6.00	
additional charges	.60	
	6.60	
holiday/sick pay @ 6%	.40	7.00
Total		13.83

Unit Labour Cost

We now know that the total labour cost to the firm of one day's work for this small gang is $13.83, and in return for this cost the physical work done should be 8 sq. metres of block wall. By simple division we can calculate that, if the labour cost to construct 8 sq. metres of wall amounts to $13.83, then the unit labour cost per sq. metres will be ($\frac{13.83}{8}$) or $1.73.

Plant, Tools and Equipment

When a substantial amount of plant is involved in carrying out an operation, it is necessary to cost it out fully. On heavily mechanised sites there will be concrete mixers, dumpers, hoists, steel scaffolding, etc and the cost of this (whether hired or owned by the contractor) will have to be recovered from the unit rates set against bill items. In this case we will assume that a mechanical mixer is not used for the mortar, so the only costs will be for scaffolding and minor tools. This can be estimated on a percentage basis, using data from past contracts, and we will take 20 per cent of the unit labour cost in this example which works out to 34 cents per sq. metre.

Materials

The next sub-heading is materials, and we have to start by working out how many blocks will go into each square metre of finished wall. The blocks must of course be 190 mm thick as specified, and we will assume that the other dimensions are 200 mm x 400 mm. One way of finding out how many of these go into a square metre would be to draw a little diagram as follows:

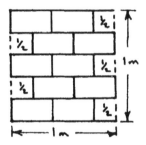

Counting up the individual blocks, the total comes to 12½ (ignoring the area taken up by mortar between the blocks). A quicker way of working this out is to first calculate the area of an individual block (0.2 m x 0.4 m = 0.08 m^2) and then divide the result into 1 m^2. The result is naturally once again 12½ blocks to the square metre.

Wastage

Unfortunately wastage of various kinds is always a factor on even the best-regulated building site, as blocks can be broken during unloading or handling, cutting is sometimes necessary to suit the drawings and theft of items such as blocks and cement may occur in some areas. Wastage comes straight out of the builder's profit margin and the sensible contractor should always be on the look-out for signs of unnecessary waste, but even so some wastage will inevitably occur and will have to be allowed for in the contractor's estimates. It is usually calculated on a percentage allowance basis, and the actual percentage will vary with the type of item and the efficiency of the contractor. In this example we will make an allowance for wastage of 5 per cent, which works out to $(12.5 \times \frac{5}{100})$ or 0.6 blocks per square metre. Thus we will base our materials cost calculation on 13.1 blocks per square metre (and this figure can also be used for ordering purposes if we get the job).

Cost per Block

Let us assume that the blocks are to be bought from a manufacturer in the nearest large town, and the price 'ex works' is 80 cents per block. This is not the figure we should use for estimating, however, since it will cost the contractor money to get the blocks to the site. Transport and handling of materials are sources of considerable 'hidden' costs to most contractors, and working out the costs in detail is the first step to an understanding of how large they can be.

Transport Cost

The cost of transport depends to some extent on whether a lorry must be hired or the contractor has a vehicle of his own. Hired vehicles are usually charged out on hourly or daily rates or a rate related to the distance covered. The contractor should also be aware of the unit costs of running his own plant and transport, and this can be done conveniently by working out his own private system of internal hire charges, so that the cost of his plant and transport can be directly recovered from contracts in accordance with the demands made upon it. It is also good discipline from foremen and site agents to appreciate that plant is expensive to purchase, run and maintain. When this starts to show up in their regular site cost returns, they find ways of using it more intensively and efficiently so that they can get it 'off their books' and back into the central yard as quickly as possible.

Calculating a Rate

In calculating a rate for plant and transport, it is vital to remember to include all cost elements including depreciation, insurance, taxes and maintenance as well as directly attributable costs such as fuel and driver's wages. The choice between an hourly/daily rate or a rate per mile or kilometre travelled will depend on the circumstances of the firm. With a daily rate, the vehicle might be used for unnecessary journeys which will push up costs for the firm as a whole without being reflected in extra output on the site. With a rate related to the distance travelled, there is the opposite danger that the vehicle may be kept on the site but only used occasionally, while there are other sites that are crying out for it and could use it more effectively. These sort of problems are best resolved by good management, but an effective compromise can be to use a distance rate underpinned by a minimum daily mileage that will be charged for in any event.

Back to the Example

Returning to the example, let us suppose that the rate per kilometre for our example firm is 60 cents. If the distance from the supplier to the site is 40 kilometres, the round trip will work out at 80 kilometres (this could be reduced if there was a reasonable prospect of a 'return load'). The only other thing we need to know is the number of blocks that the vehicle can carry — let us say 350.

Calculating the Cost

The cost of transporting the complete lorryload of 350 blocks is:

$$80 \times 60 \text{ cents} = \$48.00$$

By simple division we can calculate the cost of transporting one block as:

$$\left(\frac{\$48.00}{350}\right) = 13.7 \text{ cents}$$

Thus the cost of each block by the time it reaches the site has risen from 80 cents to 94 cents, and it is this latter figure that we should use in estimating the cost of materials.

Cost per Square Metre

We calculated that, allowing for wastage, there will be 13.1 blocks to each square metre so, by simple multiplication, the cost of blocks per square metre is:

$$13.1 \times 0.94 \qquad\qquad = \$12.31$$

Mortar

We also need to know the cost of the mortar that will be used to construct the wall. First we have to calculate the amount of mortar required for each block. Ignoring end blocks for simplicity, we know that every time a new block is laid it will be coated with mortar on two faces as shown in the diagram below:

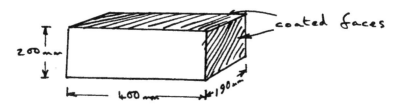

The areas of the two faces are:

0.4 x 0.19	= 0.076
0.2 x 0.19	= 0.038
Total	0.114

If the thickness of the mortar is 25 mm and there are 12.5 blocks per square metre of finished wall, the quantity of mortar required for each square metre of wall will be:

$$0.114 \times 0.025 \times 12.5 \qquad\qquad = 0.036 \text{ m}^3$$

If a cubic metre of mortar costs $50 (this too can be calculated by breaking the mortar down into its individual cost components), the cost of mortar per square metre of finished wall is calculated by simple multiplication as follows:

$$0.036 \times \$50 \qquad\qquad = \$1.80$$

Total Materials Cost

Thus the total cost of materials per square metre is:

$$\$ (12.31 + 1.80) \qquad\qquad = \$14.11$$

Total Direct Cost

We have now completed the calculation of direct costs as:

Labour	1.73
Tools and equipment	.34
Materials	14.11
Total	$16.18 per square metre

Indirect Costs

We now have to make the rather more difficult forecast of an appropriate addition to the direct site costs to cover the "indirect" costs of overheads and profit. Overheads will include site overheads (except where they are covered in the bill for "preliminaries") and general office overheads. The proportion of general office overheads to be allocated to a particular contract is usually decided on a turnover basis. In other words, if a contractor has central overheads of $25,000 per year on an annual turnover of $250,000, this means that overheads are running at 10 per cent of turnover. So a contract valued at $100,000 would be expected to contribute a sum of $10,000 to central overheads. The percentage to be allowed for profit will depend on the contractor's assessment of the risk associated with the work, as well as the general competitive situation and the size of his order book.

Site Overheads

The very experienced contractor may be able to make a fairly accurate estimate of the site overheads that will be necessary to run a particular contract on a percentage basis. But short cuts can be traps for the unwary, and most contractors would be better advised to go through the various requirements item by item and cost them out separately. This will ensure that nothing is forgotten, and will also provide a useful site overheads cost budget as a guideline for the site agent on his expenditure limits for each item.

Checklist

The actual sums involved will vary from contract to contract, but the contractor will find it useful to work out a checklist of the most common items involved to save producing a fresh list every time he prices a contract. Some of the items on a typical contract would be:

— Salary for Foreman/Agent.
— Site office.
— Office for Clerk of Works.
— Sheds and stores for materials and equipment.
— Accommodation for workmen.
— Latrines.
— Fencing to protect the site.
— Nameboard.
— Warning signs and danger lamps.
— Wages for watchmen and site services personnel.

- Site clearance and setting up costs.
- Costs of clearance on completion.
- Provision of water and other utilities.
- Insurance.

Preliminaries or Percentage?

Even where there is a separate bill for preliminaries, the contractor may choose to recover part of his site overheads by a small percentage addition to other bill items. The important point is to ensure that nothing that generates a cost is forgotten and that the costings are as realistic as possible so that these expenses are fully recovered. A further point to note is that most of these site overheads are directly related to the length of the contract — weekly wages and salaries are directly related and offices, sheds, etc should be charged out by the contractor to the job at a definite "weekly internal hire rate".

Time is Money

The fact that many site overheads are directly related to the duration of the contract means that, for the contractor: time is money. If he can manage to finish the contract early, by efficient planning and execution, he will directly save money on site overheads and that will come straight through to him as increased profit. Equally, if his performance is bad and the contract overruns, he will lose money as increased site overheads will be incurred and will not be reimbursed by the client.

General Office Overheads

General office overheads consist of all the items of expenditure that cannot be directly related to the cost of running any one particular contract, but which are necessary to keep the firm in business. They will include a salary for the contractor himself, which should be equivalent to the level he could command if he were working for an employer and is separate from the allowance for profit. Many small contractors work from a room in their house, until their firm has grown large enough to generate enough income to cover rent for a separate office. If the contractor uses a car or a pick-up to run his business, the cost of the vehicle (including depreciation) will also have to be recovered from general office overheads.

Strict Control

As with site overheads, general office overheads vary a

great deal from firm to firm depending on the size and type of the business. They need to be kept under strict control, because they can only be recovered from the physical work done on the contracts that are obtained — and these may be only one in ten of the jobs that are tendered for. Another factor is that they tend to go on being incurred whether there are contracts going on or not, so, if the intake of work slows down and turnover drops, they can only be fully recovered by increasing the percentage charge on the contracts that are obtained. This can push up unit rates, estimates and tenders — with the result that the firm becomes even less competitive!

An Annual Budget

Thus it is wise to set an annual budget covering all the items that go into general overheads, and this "general overheads budget" should be related to a "sales budget" which represents a realistic forecast of the value of work to be completed in the coming year. The sales budget can then be broken down into monthly targets (which need not be equal if the pattern of work availability is irregular due to climatic variations or the exigencies of the clients' budgetary processes). If actual work obtained drops below these monthly targets, the manager should give urgent consideration to ways of saving on general overheads so that the percentage burden on future estimates is not increased.

Checklist

The first step in working out a budget for general overheads is to set down all the relevant items of expenditure on a checklist, to ensure that nothing is forgotten. For a typical small contractor, some of the main items would be:

— Contractor's personal salary (plus oncosts).
— Salaries for office staff (clerk, secretary, etc.).
— Expenses for car or pick-up (including allowance for depreciation).
— Rent for office (and stores/workshop is applicable).
— Allowance for maintenance of buildings.
— Water, electricity charges.
— Insurance.
— Postage and telephone charges.
— Bank charges, interest on mortgages and loans.
— Audit and accountancy charges.
— Fees to private estimators (if applicable).
— Office equipment.

— Stationery.
— Petty cash items.

Example

If the total for general overheads comes to $10,000 and the forecast cost of work done by the contractor for the year is $120,000, this means that the total of direct costs incurred on the various sites is likely to be $110,000. The way to recover this is to make sure that every time $110 of direct work is done on the site an additional charge to the client of $10 is made to cover general overheads. In practice this can be achieved by a percentage addition to all net contract costs.

Net Contract Costs

The term "net contract costs" means the value of actual work done; i.e. the total contract value *less* preliminaries, prime cost sums, contingencies, etc which might not be incurred or which cannot bear an addition for overheads and profit. Thus a contract for $11,000 should recover $1,000 towards general overheads. But if $2,000 of that $11,000 relates to preliminaries, etc, the $1,000 will have to be recovered from the remaining $9,000 of billed contract items. In percentage terms, this works out at an addition of 11 per cent.

Site Overheads and Profit

We will assume that the contractor has calculated his probable level of site overheads and decided to recover 5 per cent of that from bill items. There also has to be a percentage addition for profit, as a return on capital and a reward to the contractor for bearing the risk and providing his expertise and experience. The considerations that go into the decision on the percentage addition for profit are discussed later in the book, but let us suppose in this case the contractor has aimed for a profit margin of 8 per cent.

Percentage Addition to Unit Rates

We now have three elements which together make up the percentage addition for recovery of overheads and profit from unit rates:

	per cent
General overheads	11
Site overheads	5
Profit	8
Total	24

Rate for Block Wall

This percentage addition must be applied to all direct costs in unit rates, so the final calculation for the earlier example of a rate for concrete block walls will be:

Direct costs	$16.18 per square metre
Add Overheads and Profit (24%)	3.88
Total Rate	$20.06 per square metre

Completing the Item

We can now fill in the unit rate in the bill of quantities and complete the item by multiplying the rate by the quantity as follows:

Description	Unit	Quantity	Rate	Total
Walls 190 mm thick concrete blocks bedded in mortar as specified	M^2	220	20.06	441.32

Further example

The above example is intended to serve as a gentle introduction to the craft of estimating, and later chapters will include a further example in which we shall prepare ('take-off') a bill of quantities for a simple building, and go on to produce an estimate and bid based on calculated unit rates. The author hopes to demonstrate that any competent and cost-conscious builder can prepare a satisfactory tender, providing he works through the exercise steadily step-by-step and uses his knowledge of the cost elements that he faces in operating his business.

Purchasing

Understanding costs — and always being on watch for ways of cutting them — is basic to business efficiency. The next chapter on Purchasing discusses those elements of cost which relate to bought-in items. We will then return to the general theme of estimating and bidding in the following three chapters, which will culminate in the preparation of a complete tender for a simple building.

Chapter Four

Purchasing

How much do we buy?

On an average building contract, well over half of the contractor's total expenditure goes to his suppliers of materials and components and to his sub-contractors. Much of this goes in small individual amounts from day to day and week to week during the course of the year. This is the reason why so few contractors have a clear idea of just how much they spend on these items during the course of the year, and they just note the total at the end of the financial year when the accountants add up the figures and produce a profit and loss account. This is the nearest they get to asking — and answering — the crucial question: 'How much do we buy'.

Does it matter?

For most contractors, purchasing is not by any means the most exciting part of building. It is a matter of getting out lists, reading quotations and book-keeping — more of a job for a clerk than a manager! Are they justified in shrugging their shoulders and leaving the mechanics of purchasing to a junior employee? Good managers show their skill by their ability to cut costs, and logic suggests that the biggest savings are to be found where the biggest expenditures occur. It would be a worthwhile exercise for most managers to go through their purchase ledgers to find out just how much they are spending — and what they have got in return.

Look at the figures

Let us examine the figures for a typical small general building contractor. Let us suppose that in the course of the year he has completed work to a value of $40,000 and has worked on a profit margin of 10 per cent. According to his purchase ledger, he has spent $20,000 (50% of turnover) on materials purchases and $4,000 (10% of turnover) on sub-contractors. Then let us further suppose that, by the application of more scientific and stringent purchasing techniques, it would be possible to make a saving on these costs of 5 cents in the dollar — not too ambitious a target!

	Profit ($)
$40,000 contracts completed at 10% profit margin	4,000
Possible additional saving on $24,000 purchases (5%)	1,200
Possible profit	5,200

If the firm was to achieve a profit of $5,200 using its present approach, it would have to increase its turnover by $12,000 or 30 per cent. In other words a mere 5 per cent saving in the cost of buying in outside supplies and services (without any improvement in internal operational efficiency) has produced as good a financial position in profit terms as would be achieved if the contractor increased his workload — and his worries! — by almost one-third.

Can it be done?
The above prospect sounds very attractive — but is it easier said than done? Most contractors believe they are already buying at the keenest possible prices, and cannot believe that further savings could really be obtained. Mostly they are wrong, and this chapter aims to review some of the things to look out for in purchasing building materials and sub-contractor services in the hope that the reader will be able to save money for his firm by getting these costs under control.

Control
Control is the key word — for a building business out of control, like a car out of control, can do great damage to its occupants — and the people around it. The main differences are that the damage is financial rather than physical and it shows up rather more slowly. Since material costs usually take the lion's share of expenditure, they must be brought under control if the business as a whole is to be controlled.

Cost and Price
There is a difference between cost and price, which is lost to contractors who fail to appreciate the opportunities for *cost-cutting* and *profit maximisation* through effective purchasing procedures. They should understand that buying at the lowest *price* does not automatically lead to the lowest possible *costs*. Among the factors which must be added by the contractor to the cash price (net of discounts) that he pays to the supplier to give the *total real costs to him* are:

— purchasing overheads (contractor's office costs)
— costs incurred in making good defective materials

- costs of carrying stocks
- losses due to interrupted production on the site
- costs of double handling of materials
- costs of poor service by the material supplier (or the sub-contractor)

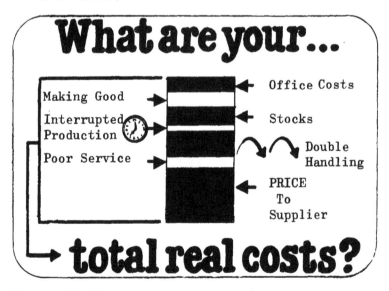

Cost and Profit

It is this *total real cost* that we must aim to cut if we are to boost profits. Of course, in some cases, it happens that the cheapest supplier is also the one who can supply materials, components or services — on the site — at the lowest overall cost. But sometimes, poor performance on the other six 'hidden' cost elements far outweights the benefit of a slightly lower offer price. It is the function of the purchasing manager (or in most small firms, the general manager, for he does this job too), to look carefully at all the cost elements — including price — and cut real costs so he can provide real profits.

Six Questions

1. Who is in charge of purchasing — is he a real professional in the sense that he really knows his job?
2. How much time do site staff give to controlling material costs (through cutting out wastage, avoiding double handling, etc.) as against controlling labour costs?
3. How much effort goes into controlling material costs in the organisation as a whole?
4. What is the volume and cost of 'emergency' last minute purchases from the site compared to planned purchases from the centre?
5. What proportion of deliveries come incomplete — or late -- or both? Do you identify and, if necessary, blacklist offending suppliers?
6. How much wastage (scrapping useless or spoiled materials plus pilferage) occurs on your sites? Who is to blame?

The Answers

The reader must answer the above questions for himself. He might as well answer them honestly, because no one else will read them and there is little point in fooling oneself! If you have the answers — even if they are unsatisfactory — you are already ahead of the general run of contractors. You have started to identify your problems, which is always the daunting first step along the road to finding a solution.

All Right — or All Wrong?

Whenever, and whatever, he is buying — the contractor should be aiming to purchase:

- the right quality
- at the right time
- in the right quantity
- from the right source
- at the right price

The first consideration in purchasing is to ensure that all goods and materials purchased are of proper *quality* and conform in every way to the specification. Most contractors

Quality Inspection

have experienced the embarrassment and difficulties caused by the replacement of defective materials condemned by the architect, and also know that materials suppliers will very rarely contribute to the costs incurred.

Time is money

Time is money for the contractor. Delays result in men and plant standing idle, while their costs go on mounting up. Contracts that are completed ahead of schedule usually show good profits. Contracts that run late usually lose money directly for the contractor, besides losing the client's goodwill and possibly incurring liquidated damages. Usually the site management takes the blame when contracts lag behind programme, but they are not always to blame. If they have to spend half their time trying to get materials delivered as promised from incompetent suppliers, there will be no time left to get the job running smoothly.

Late deliveries are the biggest problem, because they hold up the work. But early deliveries can also cause problems by congesting the site, leading to double- or even to treble-handling of materials. Every time a brick or block is moved, its real cost increases — even if this does not show up on site cost returns.

Handling Costs — The Journey of a Block

To see the costs of transport and handling, let us follow the journey of an imaginary block — and put a cost on it at every stage. Its story starts when the contractor calls for quotations and accepts the lowest offer — perhaps 40 cents per block ex works. The contractor has to pay for transport, and the cost of transporting this one block is the cost of the vehicle and driver, etc. divided by the number of blocks (as discussed in Chapter Three) — say 10 cents per block. At this stage we might note that transport from a closer supplier might have cost only 5 cents per block so, if such a supplier had offered a price below 45 cents, we have bought at the cheapest *price* but not at the cheapest *cost*.

Our block has now got to the site, but it is still in the lorry — so it has to be unloaded! Suppose the unloading of 300 blocks takes 1½ hours for 4 labourers, each of whom costs the firm 50 cents/hour. Thus the unloading costs $3.00 and the cost of our typical block has gone up by 1 cent more to 51 cents.

Later in the week the foreman discovers he has unloaded the blocks on the site of the next part of the building to be constructed — he hadn't had time to do the setting out in advance! Moving the 300 blocks takes our 4 labourers 3 hours, and the cost of our block rises a further 2 cents to 53 cents.

Because the blocks were ordered so far in advance to take advantage of the 'cheap' quotation, two more double handling exercises occur and each costs an extra 2 cents, bringing our block up to 57 cents. At long last it is now to be incorporated in the building, but its last resting place was so far from where it is to be used that it will cost a further 4 cents to get it into the hands of the mason who is to lay it.

Our '40 cent block' is now a 61 cent block apparently. But is its cost really even as low as 61 cents? Not really, because with all the movements around the site 30 of the original load of 300 blocks have been damaged and condemned by the architect. Another 20 have, unknown to the site management, been stolen by a dishonest labourer who is extending his private house. So only 250 of the original 300 blocks were actually usable. Our block was one of the lucky survivors, but it must bear its share of the cost of wastage. This is calculated by multiplying our 61 cent figure by a factor of $(\frac{300}{250})$ or 1.2 as follows:

61 x 1.2 = 73 cents

So the true cost of the '40 cent block' by the time it gets into the mason's hands is no less than 73 cents, and it is this cost of 73 cents that we must try to cut if we are to improve the contractor's profitability.

More Cost Elements Still!

The above example does not include all the possible cost elements. If materials are delivered much before they are required, there will be a long time to wait between paying the supplier and getting paid by the client. This will squeeze the firm's cash flow and may lead to interest charges if the contractor has to borrow on overdraft terms from the bank. Also a store full of materials has to be guarded to prevent pilferage, and the cost of watchmen and security fencing could be another element in the overall real costs of materials.

The Cost of Unreliability

In an ideal world, where all suppliers turned up with just what we wanted at the moment when it was to be incorporated into the structure, we would need no stores of materials because we would know that there would never be a hold-up. We do not live in an ideal world of course, and it is important to remember that unreliability is not just a nuisance — it actually costs money! A 5 per cent cheaper quotation by a stockholder for steel reinforcing bars gives a paper saving that will be wiped out in real life if the concreting gang and a hired concrete mixer are kept waiting for a week.

Paying for Reliability

If unreliable suppliers cost money, by inference we can also say — up to a point — that it makes sense to pay a little more for goods or services where there is a virtual guarantee of reliability and everything will happen as promised and exactly at the time promised. The extra price will be covered by more efficient operations on the site and the removal of the worry about missing the target date for completion. Architects are becoming more reluctant to accept late delivery of materials as an excuse for failure to meet completion dates, and heavy liquidated damages can easily be incurred.

If it has been accurately prepared, the bill of quantities will give useful guidance on how much of each type of material is to be ordered. If no bill of quantities was provided, the estimate itself must give guidance on quantities. This is just one reason why estimates must be carefully prepared — and kept — because it is much better to spend a little time taking off quantities accurately *once,* than to keep guessing

The Right Quantity

at figures but never get them quite right. For certain materials there will have to be an allowance for wastage. For example, no contractor can avoid part of his stockpile of aggregates becoming contaminated — and it would be foolish to have the job held up in its final stages while the foreman goes off in a pick-up to buy an odd box of nails.

But the contractor will not want to make too large an allowance for wastage, as it encourages site staff to be careless. The best answer is to set target percentages for wastage for all items (for expensive units like steel window frames it should obviously be *nil*) and then keep a check to make sure they are not exceeded. Some foremen are better 'housekeepers' than others — so compare the performance of all the other sites against the best — and demand to know why they don't do so well. It does no harm for the boss to show that he can recognise waste when he sees it — even to the extent of coming back to the site office with a handful of nails picked up from around the site where they had been left by employees who 'couldn't be bothered' to pick them up'.

The right source for any item or service is the one giving the best service overall at the least real cost. We say *source* rather than *supplier* because there is often a choice to be

The Right Source.

made between buying in from an outside supplier or using directly employed labour and facilities. For example, concrete blocks could be bought in or manufactured on the site. Plumbing work could be done by setting up our own plumbing department within the firm or by relying on specialist sub-contractors.

Employ or Buy-In?

The question of whether to employ 'in house' facilities or use the services of an outside supplier has to be decided on the merits of the individual case, but in general it does not pay to have one's own facilities unless they can be continuously employed. But the contractor will certainly require a reason-able minimum of his own staff, tools and equipment in order to satisfy his client that he is a real contractor rather than some kind of agent who takes a cut of the profit with-out making any tangible contribution to getting the work completed.

Price plus Service

The best supplier is the one who offers a competitive price *plus* good service. This means that the purchasing decision is not just a job for a clerk who can tell which of a series of figures is the lowest! Just as the wise client sometimes rejects the lowest tender if it comes from an inexperienced or incompetent contractor, the wise contractor must also sometimes have the self-discipline to say no.

Know Your Suppliers

It pays the contractor to really get to know his key suppliers — how they work, what are their problems and what are their strengths and weaknesses. Naturally you don't need to know *all* your suppliers so well; it doesn't matter too much where you buy your ball point pens! To make your 'getting to know your suppliers' campaign possible, concentrate on those few suppliers whose performance is really crucial to your business.

How to Choose: Pareto's Law

In making the decision on which suppliers to investigate, we are helped by the application of Pareto's Law. This states that when there are a great many sets of things to look at, it is most likely that one-fifth of them will cover four-fifths of the information. So it is likely that 20 per cent of your suppliers account for about 80 per cent of what you purchase. If you concentrate on getting a better service out of them, it will have a real impact on the running of your business and the remaining four-fifths of your suppliers can wait till you have more time.

The Top Ten

You might even decide to work out a top ten list; not of popular records but of your most 'popular' suppliers and your most 'popular' materials. As with the records, the degree of popularity is measured by the amount of money spent on them. These top ten suppliers and top ten materials are the ones where the biggest savings can and should be made.

Purchasing Power

Your top ten suppliers matter to you. Its also worth remembering that, if you are running a reasonable-sized business, you also matter to them. If they were to lose your patronage, they would lose a significant slice of their turn-over — and the profits that go with it. Your smaller suppliers, who are outside your personal 'top ten', won't care so much if you take your business elsewhere. When you are in a position to place big orders, you have *purchasing power.* It is quite legitimate to exploit this power in a calculated way to get the best possible deal for your firm. Indeed it can pay the small firm to deliberately concentrate all buying with one or two key suppliers who are known to be reliable. In this way the small firm has the same purchasing power *with those suppliers* as its larger competitors who spread their buying all around the town.

Are You a Good Customer?

'Do as you would be done by' is not a bad dictum. The contractor likes clients who know what they want, only criticise and complain when the contractor is genuinely at fault and — above all — who make payments promptly when they are due. If the contractor, in his role as customer, treats his supplier in the same way — he is likely to be popular and regarded as 'worth looking after'.

Pre-packaged Loads

One way in which a builders' merchant can help a contractor is by supplying materials in pre-packaged loads so that they are in a more suitable form for using in the structure. Timber merchants (usually for a small extra charge) are sometimes prepared to supply timber cut to the exact lengths required to save wastage due to lost offcuts on the site. In housing construction, it may pay to ask the merchant to pre-package all the ironmongery for each house unit separately, and this can also save time and therefore money on the site. A valued customer with 'purchasing power' can demand and get these extra services, which again show that there is more to good purchasing than thumbing through the quotations and picking the lowest price.

How Many Suppliers?

The general answer to the question 'how many suppliers?' is 'as few as possible', so as to maximise purchasing power. But in some areas certain key materials, such as cement, are often in short supply and a contractor who relies too heavily on one supplier can lose out if that merchant runs out of stock. In cases like this, it might pay to have 'a second string to your bow' and give a proportion of your custom (say one-third) to a second supplier, so that there is someone else to turn to in an emergency.

Competition

In some cases the element of competition between the two suppliers may give them both the incentive to try harder to please. The second supplier may adopt the slogan of the Avis Rent-a-Car company and take the attitude that 'we're number two — we try harder!' in the hope of getting a bigger slice of your business. This could lead the main supplier to offer a keener service to hold off the challenge. In other cases the main supplier may be so good that no such incentive is needed, and you can save the inevitable complications that result from using two sources of supply.

Paperwork

Another sign of a good supplier is that his paperwork (delivery notes, invoices, statements, etc.) is accurate. Dishonest suppliers sometimes invoice for goods that have not been sent or for greater quantities than were in fact supplied. Incompetent suppliers simply get their paperwork into a complete muddle and the contractor, or his office staff, consequently have to waste a great deal of valuable time sorting things out. Once again time is money, and a supplier saves both for a contractor if his paperwork is clearly set out and easy to check.

The Right Price

Last, but not of course least, comes price as a factor in purchasing. The shrewd contractor will naturally want to buy as cheaply as possible, but should always remember that the cheapest overall cost is more important than the cheapest quoted price. Sometimes, by negotiation, it is possible to get the best of both worlds by informing your regular supplier that you have had a cheaper quotation but, since you have been happy with his service, you will give him the business if he is prepared to match it. This does not always work, but it does at least show him that you are sensitive to price as well as service.

Discounts and Credit

Trade discounts are usually available to *bona-fide* builders, and these can vary greatly from merchant to merchant. The other price factor is credit, and regular customers can usually count on at least monthly credit terms provided that they are regarded as credit-worthy.

Credit-Worthiness

Most suppliers will require references before allowing a contractor to operate a credit account. Building contractors as an industry have one of the worst reputations the world over for over-trading and financial instability, so the merchant cannot be blamed for trying to safeguard himself. References will usually be required from other suppliers and sometimes also from the applicant's bank. Suppliers' credit terms can make a big difference to a contractor's cash flow, so it is well worthwhile for a contractor to make every effort — right from the start of his business career — to establish a reputation for paying his bills on time. (These aspects are discussed in more detail in the companion volume *Financial Planning for the Small Building Contractor* — Chapter Five: How to borrow money and general cash flow calculations.)

Take Your Discounts!

Credit is very valuable in helping the contractor to bridge the cash flow gap between the time that he has to pay for his resources and the dates of the interim and final payments from his client that reimburse him for his efforts. But credit itself costs money, as the supplier will expect financial recompense to make up for the time he has to wait to be paid. Thus merchants often offer quite generous discounts for prompt settlement of monthly accounts (usually 2½ per cent) and even more for cash settlement (5 or even 10 per cent). Two and a half per cent doesn't sound very much but, even if you could put off payment for another month to get extra credit without the merchant stopping supplies, it wouldn't really pay to forego the discount.

Equivalent Annual Interest

The reason why is quite simple. We are effectively paying $2.50 for $100 of credit for an extra month. By multiplying by twelve, we get an equivalent annual interest rate of no less than 30 per cent — which would put even moneylenders to shame. Given the choice between a 10 per cent cash discount and three months credit, many contractors would opt for the latter. But here they would be effectively paying 40 per cent for the money they are "borrowing".

Borrow From The Bank!

The lesson is that it is generally cheaper to borrow from the bank than from your suppliers — and every time you make a prompt payment of the sum due net of discounts, think of it as an easy way of making 30 to 40 per cent on your capital invested in the company. Even if money is tight, it pays to settle accounts on time if you possibly can, because suppliers talk to each other. If the rumour gets around that you are in financial trouble, you may find that all your suppliers cancel your credit limits and will only supply additional goods against cash settlement. This may make all the difference between weathering the storm and sinking! So remember the golden rule — settle on time and *take your discounts.*

Handling Purchasing Procedure — In The Office

Purchasing generates quite a lot of information, forms, letters and other documents; and the contractor will find that it will save a great deal of time in the long run to set up a proper system for dealing with it from the start. Once documentation gets out of control, order becomes very different to re-establish. Procedures for organising the office are discussed in greater detail in the companion volume *Accounting and Book Keeping for the Small Building Contractor* — Chapter Two: Organising the Office.

The Purchase Order

Other than minor items bought from petty cash, everything bought on behalf of the contractor should be covered by a *purchase order.* This is the starting point for each purchasing transaction, and should clearly state what — and when — the supplier is required to make available as well as where it is to be delivered (or collected). It is helpful if purchase orders are printed with the name and address of the contractor and makes for easier control if they are numbered serially in a duplicate book (or even triplicate book, so that one copy can go to the site and one remain in the office). Each order should state:

1. Name and address of supplier
2. Date
3. Quantity of order
4. Brief specification and description of quality
5. Unit price
6. Total price
7. Agreed discount (if any) and terms of payment (cash on delivery or credit period).

Book Keeping Procedures

Good book keeping procedures also save time and money in the long run, and should enable suppliers' accounts to be checked promptly as they come in and statements to be

```
┌─────────────────────────────────────────┐
│ CONTRACTOR X                              │
│ TO:  (SUPPLIER)                           │
│ DATE:                                     │
├─────────────────────────────────────────┤
│ Please supply:                            │
│                                           │
│   (ITEM, QUANTITY, QUALITY,               │
│    PRICE AND DISCOUNT)                    │
│                                           │
│                                           │
│              Signed                       │
│              (For Contractor X)           │
└─────────────────────────────────────────┘
```

EXAMPLE OF A
PURCHASE ORDER

checked and paid on the due dates. Any errors in accounts should be promptly notified to the supplier, so that a credit note for the appropriate sum can be issued to correct the mistake.

Using the Estimate

Remember to tie in your estimating and purchasing procedures. It doesn't really take much effort. When you take off your materials and price them to prepare the estimate, put the list in an envelope together with information, leaflets, etc. on any unusual materials or components that had to be investigated for pricing at that stage. If the job is not awarded to you, nothing has been lost. But if you do get it, pull out the envelope and you will find that most of the preparatory work on the purchasing side of the contract is at hand.

Handling Materials — On The Site

The aim on the site should be to cut out wastage and unnecessary work, so that materials are incorporated into the structure quickly and efficiently without undergoing the costly journey of the concrete block described earlier in this chapter. Some of the points to remember are:

Site Layout

Think about materials storage before the site is laid out, with the object of allowing for materials to be unloaded as close as possible to the place they will be needed. It helps to get the drainage and access roads completed early in the job, so that delivery lorries can be driven onto the site.

Notice of Deliveries

Never forget to warn the site foreman when a delivery is due, so that he can arrange labour for unloading and decide in advance where to store the items. Materials deliveries must be planned to ensure that they are available in advance of programme requirements and the foreman or site agent should receive a copy of all delivery schedules.

Work with the Architect

If it is not possible to order certain materials because the architect has not forwarded all the working drawings, it is *the contractor's responsibility* to ask for the details. It is not really clever to let the details get overdue, and then try to claim an extra because of the delay.

Inspection and Documentation

The contractor should designate which member (or members) of his site staff are to be responsible for checking and signing for deliveries. If breakages or faults are not found and noted *before* the load is signed for, the supplier will be quite entitled to claim that the damage occurred on the site and to refuse to issue a credit note in compensation.

Storage

Some materials, like cement, will deteriorate if they are not properly stored. Others, like clay pipes, can crack easily if they are not properly stacked. Others, such as ironmongery, may 'walk off the site' in somebody's pockets if they are not locked away in a secure shed. It is easy to lose money through giving insufficient attention to proper storage procedures.

Quality Control

The site supervisor must have sufficient practical technical knowledge to handle the materials and components for which he is responsible. For example, the mixing and placing of concrete is a much more complicated business than many small inexperienced contractors imagine. Their ignorance is costly for them — and it could well endanger the lives of the eventual occupiers of the building. This is a book about management, not technology, but here are ten simple 'commandments' for the supervisor of the concreting gang:

1. Store the cement carefully and make sure it does not become damp before it is used.
2. Ensure that the aggregate to be used is clean and check the grading with a sieve test.

3. Ensure that the water to be used is not contaminated.
4. Do not exceed the specified water/cement ratio — concretors like to add water to make the mix more 'workable': this seriously weakens the concrete and must not be allowed.
5. If the aggregates are damp, allow for this by adding less water to the mix.
6. Make sure that formwork is smooth and firmly fixed — wet concrete is heavy and collapsing formwork has led to many serious (and sometimes fatal) accidents.
7. Ensure accurate batching of every mix — batching by weight is the most accurate.
8. Ensure that every batch is thoroughly mixed.
9. Transport and place the concrete carefully to avoid segregation, and ensure that the concrete is compacted/vibrated as specified.
10. Leave the finished concrete for the full period specified before removing the formwork and props — because concrete picks up its strength quite slowly.

There are many other "points to remember" about the whole range of materials that are used in construction. The contractor should of course know them himself — but he should also pass this knowledge on to his employees. The contractor's reputation will be no better than his employees' performance so, by helping them to understand better the background to their work, he will ultimately be helping himself.

Dealing with Subcontractors

Sometimes there are real advantages to be gained by using subcontractors for specialist tasks, rather than building up and controlling one's own labour force. But it is important that the contractor should remember that, although he can delegate tasks to subcontractors, he cannot delegate his legal responsibilities under the contract. If the work done is of poor quality and is condemned by the clerk-of-works or slow progress causes the contract to overrun, it is the contractor — *not the subcontractor* — who will lose both his reputation and his profits. He can of course try to recoup some of the losses by deductions from payments to the subcontractor, but this may not cover the losses and it is not unknown for subcontractors to retaliate by walking off the site — after deliberately damaging the unfinished building, or plant and equipment belonging to the contractor. Attempting to sue a subcontractor is usually pointless, because most simply don't have enough money to pay.

Choosing a Subcontractor

Of course many subcontractors are both competent and honest, and prefer to take subcontracts rather than compete as main contractors because it involves less risk and worry. But the most competent subcontractors are not always those who will work at the cheapest piece rates. So the golden rule of purchasing still applies — look for real value by buying at the cheapest overall *cost* rather than the cheapest apparent price.

Subcontractors can't work Miracles

It is a mistake to suppose that, even if you choose a really good subcontractor, all your problems are over for that phase of the contract. Subcontractors can't work miracles — their employees are ordinary human beings just like your own staff. They may work a bit harder simply because what they earn is more closely related to what they achieve but, if there is two month's work left to complete a job and only two weeks left to the completion date, there is no way in which they can come to the rescue at the last minute.

Plan and Control

Even if most of the work on a job is let to subcontractors, the secret of success remains careful and painstaking attention to *planning and control* by competent site management. Indeed, where several different subcontractors are working on the same site, very careful scheduling of their activities is even more necessary because their personal priorities can clash — leading possibly to arguments between them and claims for delay and extras against the main contractor.

A Checklist for Subcontractor Choice

It is worthwhile to take care in choosing a subcontractor, and some contracts demand that subcontractors should be approved by the client's representative as well as by the contractor. It helps to have a checklist of questions to ask, so that nothing is forgotten.

1. Are the sub-contractor and his staff skilled craftsmen, and what kind of work are they qualified to carry out?
2. Who have they worked for before? For how long? Can they be approached for a reference?
3. Is their labour force permanent or casually-recruited?
4. What is their financial position?
5. Do they have their own tools and specialist equipment?

6. Do they have their own transport to get to the site?
7. How much work have they got currently? Is there a danger of people being switched to other jobs, leaving yours short of labour?
8. Will they co-operate with site supervisors and staff and other specialist subcontractors?

Check what you can

Many would-be subcontractors will answer the above questions with the answers they think you would like to hear. Such answers may bear little relation to the truth! So check what you can and, if time permits, try them out on a small job before entrusting them with the specialist work on a major contract.

The Subcontractor's Task

Before getting down to the final details of haggling over unit prices and stage payments, make sure that the subcontractor clearly understands his rights and responsibilities in carrying out the work. A definition of these will include:

1. A clear description of the work to be done, including responsibility for making good work of previous trades.
2. Arrangements for access and for supplies of materials and equipment.
3. The specification of the work to be done.
4. Arrangements for working area for subcontractor and storage of his tools and equipment.
5. Whether, and in what circumstances, the subcontractor may use plant belonging to the main contractor.
6. The time limits for commencing and completing the subcontract work, based on the general programme. Responsibility and penalties to be imposed if the work is delayed.
7. Arrangements for supervision and inspection by the client's representative.
8. Safety and welfare arrangements.
9. Liability for damages to third parties and insurance arrangements.

Working with the Subcontractor

Some contractors — very foolishly — look down on their subcontractors, and regard them as inferior. Remember that the subcontractor is a businessman in his own right, and ensure that the normal commercial rules of honesty and fair dealing apply. Set rates that allow him a reasonable margin for profit, or he may walk off the site in the middle of the job and leave you stranded. Make payments on time and in full — unless there are genuine grounds for a deduction for faulty work. Remember that — if he is capable — you may need him again, and his next quotation will not be so keen if he feels he has been unfairly treated on his current job.

Chapter Five

Figuring it out

The basis of estimating

The object of estimating is to produce a series of reasonably accurate predictions of the cost of the various components, resources and activities that go into the erection of a given structure on a given site within a given contract period. If all these part costs are accurate, *and providing no important cost components are forgotten,* simple addition should give an accurate prediction of the overall cost of carrying out the complete project.

How far to go?

As a general principle it can be stated that the more detailed the breakdown into cost components, the more accurate will be the final result. Thus we can make a general guess at the cost of building a standard house on the basis of its plan area and experience of constructing similar houses in the past. But we will get a more accurate idea if we put separate prices on foundations, walls, roof and services on the basis of the quantities of materials and the amount of labour that will be required for each. These cost predictions may be based on unit prices per square metre of blockwork or per cubic metre of excavation. But the cost per square metre of blockwork can itself be broken down further. We can calculate the number of blocks, the amount of mortar and the labour that is required to produce each square metre of finished block wall. Even these prices can be broken down further. The cost of blocks will be made up of the 'ex-works' purchase price, the cost of transport and handling plus an allowance for losses and breakages. The cost of mortar will include the cost of sand, cement and mixing. The estimator always faces this fundamental problem of deciding how far to go in breaking down prices into their elements in pursuit of greater accuracy.

As far as need be

The only honest answer one can give to the generalised question of 'how far to go?' is the rather unhelpful one of 'as far as need be'. At the beginning of an attempt to learn the trade of estimating, or when the job appears to have some unusual or tricky features, it is best to go right back to basics and work out all the most important quantities from their elements. Even then, some short cuts will be taken. We could, for example, break down the cost of the site sign board into the cost of timber, nails, paint and the time of carpenters, painters or signwriters and general labourers. But it is normally sufficient to allow what appears to be a reasonable figure, since an error of a few dollars either way is not very important on a 'one-off' item and it is better to concentrate the estimator's efforts on items where larger quantities and more money are at stake.

Simple arithmetic

An estimator does not need to be a graduate in advanced mathematics. All that is needed is an ability to work accurately with the basic arithmetical skills of addition, subtraction, multiplication and division plus the ability to read and understand architectural drawings. Most practical builders have these basic skills. In fact, they are essential tools of the builder's trade as the carpenter, for example, will need to be able to work from drawings and calculate the lengths of timber and numbers of nails and screws that will be required to do his part of the work. So no contractor should be afraid of estimating. It is not some peculiar science reserved for the Einsteins of this world. At the level of simple buildings, of the kind that most small contractors are in business to construct, it has much more in common with a trade. As with other trades, there are certain 'tricks of the trade' which help the experienced operator to work more quickly and effectively. But also, as with other trades, the route to success requires painstaking and persistent effort and this is best achieved by practice and yet more practice.

Know your costs!

The successful builder knows the cost of building, just as the successful baker knows the cost of baking bread. The contractor is in a dangerous position if he has to always rely on an outside private estimator for the facts and figures he needs for his tenders. Even if, through lack of time, he has to rely on someone else to do the preparatory work, he still requires sufficient knowledge to review their work and make sure that the suggested unit costs can be worked to and leave an adequate margin of profit on the completed tender. The cost estimates will also provide a valuable set of targets when he comes to manage the contract. If the estimated cost of cement was $5 per bag, then he knows that he must keep within this figure if he is to make his required profit on concreting. If the estimated cost of hand excavation is based on an output of 2 cubic metres per man day, then he will be losing money on a 1 metre wide x 1 metre deep trench if a gang of 5 labourers fail to complete a total length of 50 metres at the end of a 5-day working week. Without these cost targets, the contractor is literally working in the dark with no idea of what he is doing well and where he most needs to improve his performance.

The Cost Mix

An example of the need to be cost-conscious is concrete. When we talk of 1:2:4 concrete or 1:3:6 concrete, most builders know that this refers to the quantities of cement, sand and aggregate that form the ingredients of the chosen mix. This is the 'quantity mix', and the contractor as technologist needs to know it. But the contractor in his other role as manager/businessman also needs to know the 'cost mix' of concrete. The cost per ton of cement is often more than ten times the cost per ton of aggregate, and the cost ratio for cement to sand is even greater. Suppose that the costs per cubic metre of cement, sand and aggregate are $64, $4 and $6 respectively. Ignoring the volume of cement in the finished mix and allowing for 20 per cent shrinkage, the cost of materials in the concrete would be as follows:

1 m^3 cement	@ $64	64
2 m^3 sand	@ $4	8
4 m^3 aggregate	@ $6	24
		96
Add 25% to allow for 20% shrinkage		24
Cost for 6 m^3 concrete		120
Thus cost per cubic metre =		$20

The cost ratio for the three ingredients is 64:8:24 or 8:1:3. So we could say, from a cost mix point of view, 1:2:4 concrete by volume of cement:sand:aggregate is 8:1:3 by cost. The relationship is illustrated in the diagram opposite.

Concentrate on high cost items

Once the high cost ingredients and components have been identified, the manager knows where to concentrate his attention in order to achieve the maximum savings. Quite clearly we would gain more by negotiating a 10 per cent

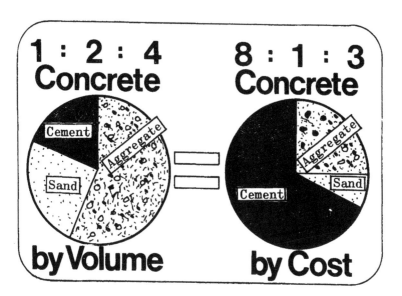

1 : 2 : 4
Concrete

8 : 1 : 3
Concrete

Cement

Aggregate

Sand

Aggregate

Cement

Sand

by Volume

by Cost

discount on cement purchases than we would from one of 20 per cent on sand or aggregates. Equally cutting wastage and losses of cement by a tenth would save more than cutting wastage and losses of the other ingredients by a fifth. Naturally all savings are worth having, but the time of every manager and superintendent is limited. Focusing attention on high cost items is the best way to ensure that this time and effort achieves the best possible return.

The Bill of Quantities

On larger jobs it is quite common for the contractor to be provided with a bill of quantities as part of the tender documents. This will set out the work to be done in a systematic way and each item will have a quantity against it, such as a volume of concrete, a weight of steel reinforcement, an area of formwork, a length of trench or a number of manhole covers. This speeds up the tendering process for the contractor, but it is still his responsibility to check that the bill of quantities ties up correctly with the drawings.

Smaller Contracts

On smaller jobs the contractor usually only receives the drawings and specifications, so he has to figure out the quantities involved on his own in order to price the contract accurately. The ability to break down the general mass of information contained in the drawings into the simple list of facts and figures that makes up a bill of quantities is an essential skill that the contractor must learn. In this chapter

we will 'figure out' the quantities for a simple small building.
The calculations will not be complicated, providing we work
through them in a systematic step-by-step fashion. The main
thing is not to be frightened of the detail and mass of the
figures, to work accurately and neatly, and finally to check
that the total figures are more or less what we would expect
and that nothing has been left out.

The Building

The building that we shall prepare a tender for is illustrated
on the next few pages. It is a simple open structure 10 metres
long by 6 metres wide, and would be suitable as a school
classroom or a small workshop. There are double access doors

This is
THE BUILDING . . .

for which we shall
prepare a tender

ELEVATIONS –

SOUTH

EAST

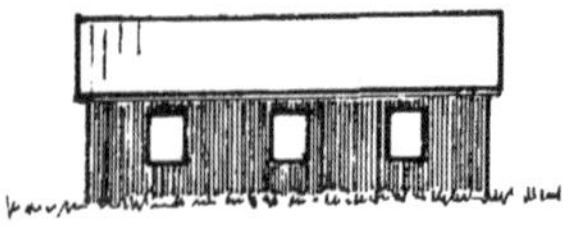

NORTH

WEST

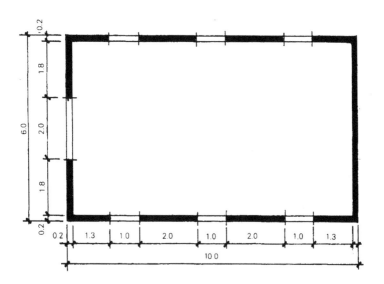

PLAN OF BUILDING (scale 1:100)

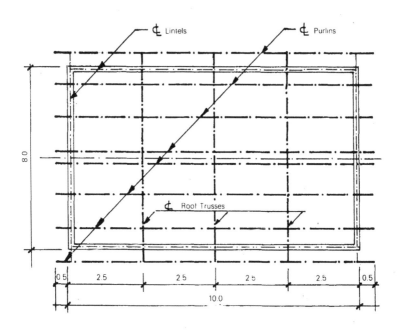

PLAN OF ROOF (scale 1:100)

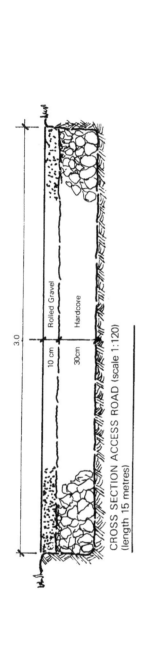

CROSS SECTION ACCESS ROAD (scale 1:120)
(length 15 metres)

3.0

10 cm — Rolled Gravel

30cm — Hardcore

GABLE WALL (West) (1:100)

Concrete Blocks

Concr

20 x 25 x 240 cm

5.2

1.5   2.5   1.0   0.2

SIDE WALLS (1:100)

Concrete Lintels 20 x 25 x 140 cm

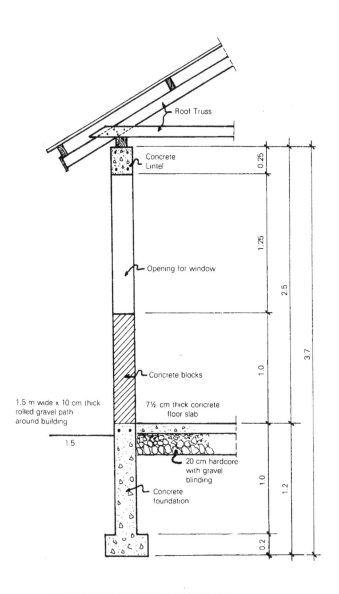

Roof Truss

Concrete
Lintel

0.25

1.25

2.5

Opening for window

3.7

Concrete blocks

1.0

1.5 m wide x 10 cm thick
rolled gravel path
around building

7½ cm thick concrete
floor slab

1.5

20 cm hardcore
with gravel
blinding

Concrete
foundation

1.0

1.2

0.2

SECTION LONGITUDINAL WALLS (1:20)

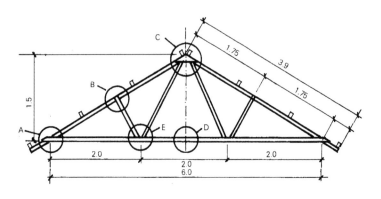

ROOF TRUSS (1:50)

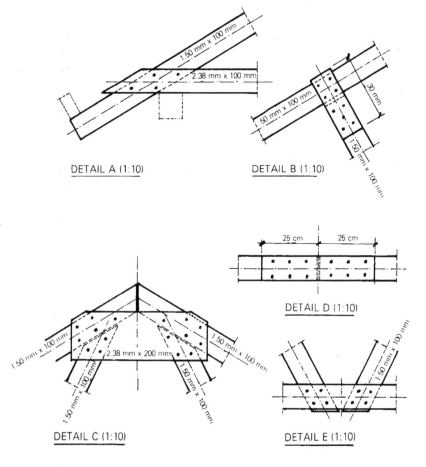

DETAIL A (1:10)

DETAIL B (1:10)

DETAIL C (1:10)

DETAIL D (1:10)

DETAIL E (1:10)

102

at one end of the building and three windows in each side. The walls are of concrete blocks, plastered both externally and internally. The roof is to be covered with corrugated iron sheets.

Access

The site is adjacent to a main road, and a 3 metre-wide hardcore and murram access road is to be constructed as part of the main contract. Ample space is available for the storage and stockpiling of plant and materials on the site.

General Provisions

We shall assume a typical specification for the work, and a fairly comfortable contract period of 3 months. Since the site is only about 10 km from the contractor's home town, transport and setting up costs will not be exceptional.

Break down into trades

The first step is to decide on how many separate bills of quantities are required. The bills can be separated in any way that appears logical, but the best way is usually to separate them according to the different trades that are involved in the work. This method helps with planning and costing, and is particularly convenient if the work in one or more of the trades is to be sub-contractred since the appropriate bill or bills can be handed over to the sub-contractors complete for pricing.

Bill No.1 — Preliminaries

The first bill is always known as preliminaries, and includes all the 'odds and ends' of expenditure that are necessary in order to get the contract underway. It will include such items as a site office, fencing, materials storage, provision of a water supply, insurance and general tools and equipment that cannot be conveniently costed out to separate bills and items. Since we are at present concentrating on figuring out the quantities for the building itself, we will not go into more detail on *Bill No.1 — Preliminaries* at this stage.

Bill No.2 — Groundworks

Bill No.2 will be groundworks and will cover site clearance, excavation, trenching, backfill, placing and compaction of hardcore and gravel filling. Although the bill of quantities is not being prepared for any other purpose than as a guide to the contractor in formulating his overall price for the work, it is still best to proceed logically with each separate activity set down as a numbered bill item, 2.01, 2.02, 2.03, etc. This logical approach will pay off when we have finished writing out the bills, and need to check that all the various activities have had bill items allocated to them so that they will all be priced. Quite clearly there is no point in going to all the trouble of working out and pricing a detailed bill of quantities if certain key activities (that will give rise to costs) are left out of account. If that were to happen, the apparent accuracy would be an illusion and the unbilled work would have to be carried out by the contractor at his own expense.

Excavation plan

Thus there is no safe substitute for the painstaking step-by-step approach, and we will start by producing a drawing that will show the areas and depths of excavation more clearly, assuming that a 0.8 m wide trench will be required in order to construct the foundation walls. We will assume a topsoil depth of 0.1 m, which will take us down to the formation level for the gravel paths around the building. We then have to go down a further 0.1 m to reach the formation level for the hardcore under the floor slab, since the concrete slab inside the building is laid above existing ground level.

Trench excavation

We have now removed two horizontal slices of soil each 0.1 m thick, and the only excavation that remains over the area of the building is that required for the foundation trenches. Since the base of the foundation is 1.1 m below existing ground level, the trench excavation will be a further 0.9 m deep.

Separate items

It is standard practice to have separate items for excavation in trenches and shallow excavation over a larger area, since the contractor will adopt different methods for the two types of excavation and these are quite likely to give rise to different costs. For example shallow excavation to a formation level including "bottoming up" will probably be carried out by hand, but a mechanical digger may be employed for trenching. A further factor is that in poor ground trenches have to be protected by timbering or even sheet piling, and this additional cost will have to be covered by the contractor in his rates for excavation.

Deeper means dearer

The trenches in this example are quite shallow, but the general rule for trenching is that 'deeper means dearer'. Thus the cost per cubic metre of excavation for a 2 metre-deep trench will be greater than for a 1 metre-deep trench, and the overall cost per metre run of trench will be *more* than twice as much. To allow the contractor to vary his unit prices appropriately, it is standard quantity surveying practice to have separate items for excavation to various depths, e.g.:

1. Excavation to 1 metre below ground level.
2. Excavation from 1 to 2 metres below ground level.
3. Excavation from 2 to 3 metres below ground level, etc.

Method of measurement

Foundation trenches are normally measured according to the number of cubic metres of excavation but trenches of specified and uniform width, such as sewerage trenches, can be measured more conveniently in terms of their length in metres.

Plan and section

A plan and section showing the excavation required for the building in our example is given below:

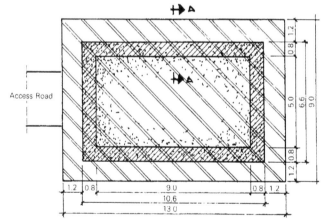

EXCAVATION PLAN (Scale: 1:100)

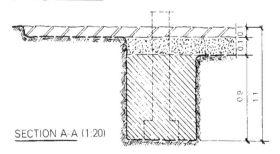

SECTION A-A (1:20)

2.01 Site clearance and removal of topsoil to approx. depth 10 cm

Before we can start on constructing the building it will be necessary to clear the area of bushes and undergrowth and then remove the topsoil and place it in a stockpile. It is usual to measure this work in square metres of ground affected. In the example, this work will be required over the area covered by the access road and the paths around the building as well as the building itself. For this item there is no need to separate out the pathways from the building, and the calculation consists quite simply of finding out the area of two rectangles as shown below:

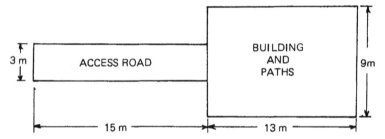

Set out in this way the problem becomes very easy, and a small sketch is often a great help to the estimator in showing the easiest method of calculating the quantity he requires. In this case the calculation is as follows:

Building and paths	13 x 9 =	117
Access road	15 x 3 =	45
	Total =	*162 sq. metres*

2.02 Excavation to lower level of hardcore over building area

Looking at the excavation plan, we note that the area to be taken down to a level of 0.2 m below ground level consists of a rectangle 10.6 metres x 6.6 metres. It could be argued that this item should cover the area of 9 metres x 5 metres within the trench area since this is the only part to be 'bottomed up', but for ease of calculation we will take the complete slice of excavation between 0.1 and 0.2 m below ground level as a single item. In this case a bulk measure in cubic metres is appropriate and the volume of material involved is:

10.6 x 6.6 x 0.1 = *7 cu. metres*

2.03 Excavation in foundation trenches

Referring back to the excavation drawings we note that

the depth of excavation is 0.9 metres and the area to be excavated is as shown below:

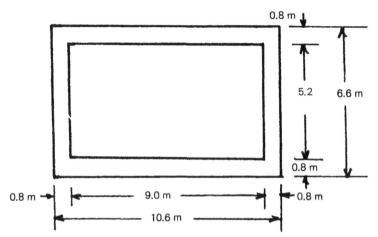

There are two ways of carrying out this calculation.

Method One. We know the width of trench is 0.8 metres. The length (remembering not to double-count the corners) is $(2 \times 10.6) + (2 \times 5.0) = 21.2 + 10.0 = 31.2$ metres.

Thus volume of excavation = $0.8 \times 0.9 \times 31.2 = $ **22.5 cu. metres.**

Method Two. The area that we need is the area that lies between two rectangles, so a short cut to the plan area of the trenches is to calculate the area of these rectangles and subtract one from the other as follows:

Plan area = $(10.6 \times 6.6) - (9.0 \times 5.0) = 70 - 45 = 25$ sq. metres.

Thus volume of excavation = $0.9 \times 25 = $ **22.5 cu. metres.**

In this case the short cut in Method Two did not save a great deal of time, but on complicated calculations the experienced estimator can save a lot of effort by always being on the look-out for short cuts that will get him the answers he needs with the minimum number of individual calculations.

2.04 Prepare bottom of trench for concreting

It will be necessary to ensure that the trench bottom is completely smooth and level over the area where concrete is to be poured. This work need not be done over the full width of the trench, and so we will need to measure the 0.8 metre wide area that will be covered in concrete. Using the 'difference between two rectangles' method again, the area is:

$(10.2 \times 6.2) - (9.4 \times 5.4)$

$= 63.24 - 50.76 = $ **12.5 sq. metres**

Yet another alternative method of calculation is to calculate the length of trench along the centre line as follows:

Length of trench = 2 x (9.8 + 5.8) = 31.2 metres
Thus area = 0.40 x 31.2 = **12.5 sq metres**

2.05 Return, fill and ram excavated material around foundation

After the foundation walls have been concreted, the trench must be backfilled in layers and well-rammed so as to provide a good support to the concrete slab inside the building and the gravel paths around the perimeter. The item will be measured in cubic metres, and will be the volume excavated (item 2.03 and part of 2.02 above) less the volume of concrete and hardcore that will intrude into this volume as indicated in a revised Section A-A below:

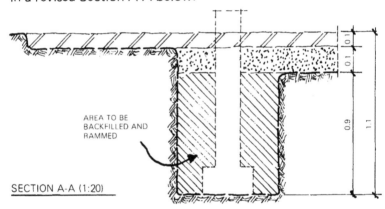

AREA TO BE
BACKFILLED AND
RAMMED

SECTION A-A (1:20)

The total excavation covered in 2.03 amounts to 22.5 cu.m. Since this item will not attract very high unit rates, an approximation is permissible and the volume to be deducted from 22.5 cu.m. can be based on the concrete wall thickness of 0.2 metres and a height of 0.9 metres. The length of trench along the centre lines was calculated for the last item as 31.2 metres.

Thus volume to be deducted = 0.2 x 0.9 x 31.2 = 5.6 cu. metres.

Thus volume of backfill = 22.5 − 5.6 = say **17 cu. metres.**

2.06 Place and compact hardcore 20 cm deep within building and blind with gravel to receive concrete

Since the *external* dimensions of the building are 10 metres by 6 metres and the foundation walls are 0.2 m thick, the internal dimensions must be 9.6 metres by 5.6 metres. We could now go straight on to calculate the volume of hardcore, but one of the 'tricks of the trade' in estimating is to carry out the calculations in such a way that partial calculations can be used for one item after another. So as a first stage we calculate the internal area of the building, since this will help us with the volumes of hardcore, gravel and concrete as well as the area to receive a floor screed after the concrete has been laid. Since this item covers both hardcore and gravel blinding, it would probably be measured in square metres in any event. The calculation is:

Area = 9.6 x 5.6 = *53.8 sq. metres.*

2.07 *Adjust levels, supply and roll gravel path 10 cu min thick around building*

Since the theoretical formation level for the gravel path is only 0.1 m below existing ground level (which will not in practice be completely level and horizontal) and 0.1 m of topsoil will have to be removed, the paths will have to be more than 0.1 m thick in places if the top of the paths is to be level all round the building. It would not be worth having a separate item to measure this extra fill, so item 2.07 will be measured in square metres but the contractor will have to make an additional allowance in his prices for the extra material and work involved. We could calculate the path area by the 'difference of two rectangles' but it is slightly quicker to base the calculation on the lengths of the paths measured along their centre lines. This means adding half the path to each of the outside dimensions of the building, i.e. 10.75 and 6.75 metres.

Thus area of paths = 1.50 x (10.75 + 6.75) x 2
= *52.5 square metres*

2.08 *Excavation for access road*

The formation level of the access road is (on average) 0.4 m below existing ground level. But item 2.01 covered the removal of 0.1 m topsoil, so an average 0.3 m remains to be removed under this item. The area of the access road was found under item 2.01 to be 45 sq. metres, so the volume of excavation for the access road is:

45 x 0.3 = *13.5 cubic metres*

2.09 *Supply and compact hardcore to access road, min. 30 cm. deep*

This item will be measured by volume and, since the depth of hardcore is 30 cm, the volume must be exactly the same as for the previous item:

13.5 cubic metres

2.10 Rolled gravel surface to access road 10 cm thick
This is a 'finishing' item and is most conveniently measured in square metres. As already calculated, the area of the access road is:

45 square metres

2.11 Level surplus soil as directed and tidy site on completion
Although it would be possible to measure the volume of surplus soil and calculate an average length of haul, this would not really be worthwhile for a minor item and would also give rise to difficulties and disputes in measurement on completion. Thus items of this kind are more conveniently dealt with as 'lump sums', in which the contractor estimates a fixed amount for carrying out the work and it is up to him to make sure that he makes full allowance for everything that has to be done. So against this item we put *LS* (lump sum).

Bill No.3 — Concrete
This bill will cover the concrete foundation base and walls and the concrete floor slab, including formwork and steel reinforcement. In this case we will, however have a separate bill for walling (including lintels) and for plastering (including the floor screed). The most convenient way of dealing with the concrete items is to group together the formwork items (measured in square metres), the concrete items (measured in cubic metres) and the steel reinforcement items (measured in kilograms).

3.01 Formwork to foundation base
From the section through the walls given in the general drawings we can sketch the cross section of the concrete foundation as shown opposite.

We will have separate formwork items for the foundation base and the foundation walls to make the example clearer, although in practice they might easily be combined. There will be no formwork item for the floor slab, since it will be supported at the sides by the foundation walls.

We know that the depth of the foundation base is a uniform 20 cm, so to get the area of formwork we need to multiply by the distance round the base on the outside plus the distance round the base on the inside. As an aid to calculate these distances we can sketch a plan view of the foundation as opposite.

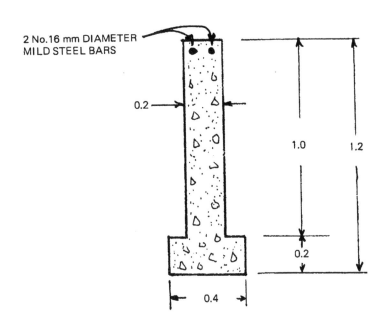

2 No.16 mm DIAMETER
MILD STEEL BARS

0.2

1.0   1.2

0.2

0.4

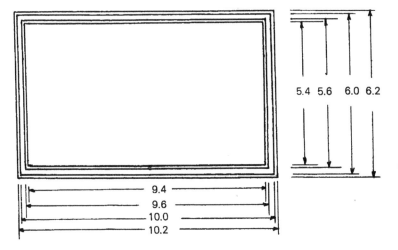

5.4 5.6   6.0 6.2

9.4
9.6
10.0
10.2

The distance round the base on the outside measures 10.2 by 6.2 metres, so the total distance is 2 × (10.2 + 6.2) or 32.8 metres. The distance round the base on the inside measures 9.4 by 5.4 metres, so the total distance is 2 × (9.4 + 5.4) or 29.6 metres. Since the depth of the base formwork is 0.2 metres, the total area of formwork required for this item must be:

0.2 × (3.8 + 29.6) = *12.5 sq. metres*

3.02 Formwork to foundation walls

Referring back to the section through the foundation, we note that the depth of the walls is exactly 1 metre. Our sketch plan shows that the walls measure 10.0 x 6.0 metres externally and 9.6 x 5.6 metres internally. Thus the total area of formwork required under this item is:

1.0 x (32.0 + 30.4) = **62.4 sq. metres**

3.03 Concrete to foundation base

The foundation base is 0.4 m wide and 0.2 m deep. Its dimensions, measured along the centre lines, are 9.8 metres by 5.8 metres. Thus the volume of concrete will be:

0.4 x 0.2 x 2 x (9.80 + 5.8) = **2.5 cubic metres**

3.04 Concrete to foundation walls

The foundation walls are 0.2 m wide and 1 metre deep. Their dimensions, measured along the centre lines, are again 9.8 by 5.8 metres.

Thus the volume of concrete in the walls will be:

0.2 x 1.00 x 2 x (9.80 + 5.80) = **6.2 cubic metres**

3.05 Concrete to floor slab

The floor slab measures 9.6 metres by 5.6 metres, and is 7.5 cm thick. Thus the volume of concrete will be:

9.6 x 5.6 x 0.075 = **4.1 cubic metres**

3.06 Steel reinforcement to foundation walls

We will assume that the foundation base is of unreinforced mass concrete, but that two 16 mm diameter mild steel bars should be included in the top of the walls running right around the building and lapped to provide continuity. Steel is generally sold by the stockholder or supplier by weight, so the unit of measurement will be kilograms. Standard tables are available giving the weights of steel reinforcement bar of various diameters per metre length, and the figure for 16 mm bar is 1.58 kg/metre.

The centreline dimensions are 9.8 metres by 5.8 metres and we will allow for two laps of 0.6 metres, so the total length of bar required is:

2 x 2 x (9.8 + 5.3 + 0.6) = 64.8 metres

Thus weight = 1.58 x 64.8 = **102 kg**

3.07 Steel reinforcement to floor slab

If mesh reinforcement was available, this would be the most suitable for this purpose since it is easier to handle and reduces steelfixing costs. However, in this case we will assume that the reinforcement is to consist of 8 mm mild steel bars laid at 30 cm centres in both directions to provide a 'mesh effect'. The floor slab measures 9.6 metres by 5.6 metres, so the lengths of the bars will be 9.5 metres and 5.5 metres respectively. Next we need to find out the number of bars that will be required. One method would be to draw a plan of the floor slab with the bars at 30 cm centres as illustrated below and simply count them:

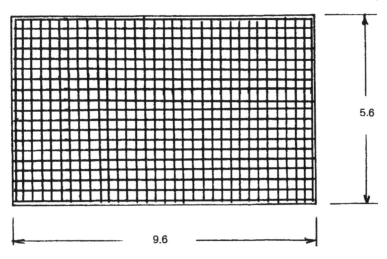

5.6

9.6

Counting the bars in both directions, we find that there are 19 of the 9.5 metre bars and 33 of the 5.5 metre bars.

But producing a drawing takes quite a long time. So is there a quicker way? Yes — the number of bars can be calculated by dividing the distance 'd' between the bars into the length 'L' to be covered and adding one. So the formula is:

Number of bars $= \frac{L}{d} + 1$

In this case, for the 9.5 metre bars: L = 5.6 and d = 0.30
Thus number of 9.5 metre bars $= \frac{5.6}{0.3} + 1 = 18 + 1 = 19$

For the 5.5 metre bars: L = 9.6 and d = 0.30
Thus number of 5.5 metre bars $= \frac{9.6}{0.3} + 1 = 32 + 1 = 33$

The total length of 8 mm bar required is:

(19 x 9.5) + (33 x 5.5) = 180 + 181 = 361 metres. es.

8 mm bar weighs 0.395 kg/metre, so the total weight of

8 mm bar required will be:

361 x 0.395 = **142 kg**

Note: Calculation of the number of bars was quite easy in this case because the floor slab was of a simple rectangular shape. Sometimes one is faced with calculating the weight of reinforcement required in slabs of irregular shape, but in which the bars are a standard distance apart. In such a case a reasonably accurate short cut is to calculate the weight of reinforcement in a typical one metre square section of the slab and multiply by the total area. If the distance between bars is 'n_1', in one direction and 'n_2' in the other direction and the weight per metre length is 'w', the weight of bars per square metre will be:

$$(\frac{1}{n_1} + \frac{1}{n_2}) \times w$$

Using the formula on our example the weight per square metre is:

$$(\frac{1}{0.3} + \frac{1}{0.3}) \times 0.395 = 2.63 \text{ kg/sq. metre}$$

Since the area is 9.6 x 5.6 metres, the total weight using this method of calculation is:

9.6 x 5.6 x 2.63 = 141 kg

Bill No.4 — Blockwork and plastering

This bill will cover items describing blockwork, lintels, internal and external plastering and the floor screed. Blockwork, plastering and screeding will be measured in square metres, but the unit for the lintels will be No. (number) since it would extend the bill unnecessarily to have separate items for formwork, concrete and reinforcement for each kind of lintel. These costs will have to be calculated or estimated when we come to price the bill.

4.01 Concrete block walls 20 cm thick

In order to get the net area of walls that will be made up of concrete blocks we must start by calculating the gross area of the side and gable walls and then deduct the area that will be taken up with the doorway, windows and lintels.

We will start with the gable walls. Referring back to the drawings we see that each gable wall consists of a rectangle 2.5 m high by 6.0 m long with a triangle 1.5 m high by 6.0 m long on top of it. Thus the gross area of the two gable walls will be:

$$2 \times [(6.0 \times 2.5) + (\tfrac{1}{2} \times 6.0 \times 1.5)]$$
$$= 2 \times (15.0 + 4.5) = 39.0 \text{ sq.m.}$$

The east gable wall is blank, but we need to deduct for the doorway and lintel over in the west wall. The calculations are:

Doorway	2.00 x 2.25	= 4.5	
Lintel	0.25 x 2.40	= 0.6	5.1
Net area of gable walls is 39.0 − 5.1 =			33.9 sq.m.

Next the side walls. Although the side elevation is 10.0 m long, we will base our calculation on a length of 9.6 m since 20 cm at each end have already been included in the gable wall calculations. Thus the gross area of the side walls is:

2 x 9.6 x 2.5	=	48.0 sq.m.

There are three windows and lintels in each elevation so we deduct:

Windows	6 x 1.00 x 1.25	= 7.5	
Lintels	6 x 0.25 x 1.40	= 2.1	9.6
Side walls less windows and lintels is			
48.0 − 9.6 =			38.4 sq.m.

Thus net area of concrete blockwork = **72.3 sq.m.**

The next two items refer to the lintels and, since the unit is to be No., we have only to count them!

4.02 Window lintels 20 x 25 x 140 cm   **6 No.**

4.03 Door lintel 20 x 25 x 240 cm   **1 No.**

We now go on to wall plastering, and will measure external and internal plastering separately.

4.04 External wall plastering, to receive paint

The calculation will be the same as for 4.01 above except that we will not deduct the area of lintels, and the side wall calculation will be based on the full width of 10.0 metres since this is a skin going on the external face of the wall rather than the wall itself.

Gable walls

Gross area: $2 \times [(6.0 \times 2.5) + (\tfrac{1}{2} \times 6.00 \times 1.50)]$
= 39.0

Less doorway: 2.00 x 2.25   = 4.5

39.0 − 4.5		= 34.5

Side walls
Gross area: 2 x 10.0 x 2.5   = 50.0
Less windows: 6 x 1.00 x 1.25 = 7.5   = 42.5
    Thus net area of external plastering
    is 34.5 + 42.5   = *77.0 sq.m.*

4.05 Internal wall plastering, to receive paint

The same principle applies to the calculation of the internal surface area, based on the internal dimensions of 9.6 metres by 5.6 metres, although the top triangles of the gable walls will of course be the same size internally as externally.

Gable walls
Gross area: 2 x [(5.6 x 2.5) + (½ x 6.0 x 1.5)]
  = 37.0
Less doorway: 2.00 x 2.25   = 4.5
37.0 − 4.5   = 32.5

Side walls
Gross area: 2 x 9.7 x 2.5 = 48.0
Less windows: 6 x 1.00 x 1.25= 7.5   = 40.5
    Thus net area of internal plastering
    is 32.5 + 40.5   = *73.0 sq.m.*

4.06 Floor screed, average thickness 3 cm

The area of floor screed must be the same as the area of the concrete floor slab (9.6 metres by 5.6 metres). Thus the quantity for this item is 9.6 x 5.6 = *53.8 sq. metres.*

4.07 Supply, position and grout in place ½" anchor bolts for roof fixing

One further item must be added before this bill is complete. It will be necessary to fix 4" x 4" timbers to the top of the side walls to support the trusses and to the gable walls to support the roof at the east and west ends of the building. This will be done by cutting out holes 6" deep in the top of the wall, placing ½" anchor bolts in position and setting them firmly in position with a 1:1 sand/cement grout. Four bolts will be required for each side of the two gables, making sixteen in all. Seven bolts will be needed in each side wall, making fourteen in total. Thus the total number will be thirty, and the final item in Bill No.4 will be:
30 No.

Bill No.5 — Carpentry and windows

These items will cover metal windows, wall plates, trusses, fascia board, the door and associated ironmongery. On more complex buildings it is sometimes necessary to provide a separate bill for ironmongery, steel windows, etc. The first item will be the windows and it is common practice to specify those of a particular manufacturer 'or similar approved', so that the unit will be No.:

5.01 Supply and fix 1.00 x 1.25 steel windows 6 No.

5.02 100 mm x 100 mm wall plates, including fixing to anchor bolts

The wall plates to the side walls will be 9.6 metres long. From the drawing of the trusses we note that the gable wall plates will measure 3.9 metres at each side. Thus the total length is:

$(2 \times 9.6) + (4 \times 3.9) = $ *35 metres*

5.03 Roof trusses as described, supply and fix

The roof trusses will have to be 'broken down' for pricing purposes, but in the bill they will appear as *3 No.*

The trusses are to be held down onto the wall plates with metal straps to provide additional security against roof damage in strong winds. One strap will be provided at each end of each truss, so six will be required in all. Thus item 5.04 will be:

5.04 No.16 gauge hoop iron strap 32 mm wide and 40 cm long including screw fixing to wall plates   6 No.

5.05 Supply and fix 50 mm x 100 mm purlins

This is a very simple calculation. Each purlin is 11 metres long, and four are required on each side. Thus the total length of 50 mm x 100 mm timber required is:

$2 \times 4 \times 11.0 = $ *88 lin.m.*

5.06 Fascia board 31 mm x 150 mm

Fascia board will be required to protect the edges of the roof at the side walls and gables, and the total length of timber will be:

$(2 \times 10.0) + (4 \times 3.9) = $ *36 metres*

Note: It is usual to 'round off' quantities for low cost items such as this. Working to decimal places is only justified for items such as concrete where unit rates are more costly.

The final item in this bill will be the wooden doors and frame to be built into the west gable wall. For the purposes of bill preparation this will be included as a single item, including ironmongery and the cost of fixing:

5.07 Double doors 2.00 x 2.25 m including frame, hinges, lock and fitting   *1 No.*

Bill No.6 — Roofing

The roof will be of galvanised corrugated mild steel sheets screwed to the timber trusses and gable supports, and capped with ridge pieces. The unit of measurement will be square metres of roof surface for the sheets and linear metres for the ridging.

6.01 Roof sheets, including laps and fixing as specified

The length of the roof will be 11.0 metres, so the total surface area is:

$2 \times 11.0 \times 3.90 =$ *85.8 sq. metres*

6.02 Ridges 38 cm girth, including fixing as specified
11 metres

Bill No.7 — Painting and glazing

In order to simplify the example there has been no provision for plumbing and electrical works, which are often sub-contracted to specialists. Thus all that remains to be done to complete the building is to glaze the windows and paint the building inside and out.

7.01 Window glazing as specified

Where windows of many different shapes and sizes are to be glazed, it is usual to calculate the area of glass required. In this case all the windows are the same size, so the unit can be the number of windows, i.e.   *6 No.*

7.02 Paint external surface of walls as specified

Since we are going to paint the area that has been plastered, the quantity must be the same as for item 4.04:   *77 sq.m.*

7.03 Paint internal surface of walls as specified

This follows item 4.05 and is therefore: *73 sq.m.*

Windows and doors will have to be painted, and the unit here can be the number to be painted:

7.04 Paint windows as specified   *6 No.*

7.05 Paint doors and frames as specified   *1 No.*

Finally we must include an item for painting the fascia boards. It is not really worthwhile to go to the trouble of calculating the area involved, so the easiest way is to mark this item as LS (lump sum) against which the contractor puts a cash price for doing this work. The main trouble with measuring items by means of 'lump sums' is that they make it more difficult to agree new prices with the client if quantities change due to an alteration in design. If, for example, the client decided (before building commenced) to increase the length of the building from 10 metres to 11 metres, items measured in metres, square metres, kilograms, etc. could be simply recalculated on the basis of increased quantities but the same unit rates. If this was thought to be a real possibility, we could measure painting of fascia boards by their total length (36 metres), but in this case we will say:

7.06 Paint fascia boards as specified   *LS*

Bill No.1 — Preliminaries

We have now prepared a complete bill of quantities covering all the materials and physical work that will go into completing the building in accordance with the design set out in the drawing and specifications. This work has been set out according to the trades and types of work involved in bills 2-7. All that remains to be covered in detail is Bill No.1 — Preliminaries, which gives the contractor an opportunity to put a price on all the miscellaneous items of expenditure that he will have to face in order to carry out the work in accordance with the general and specific conditions of contract.

The contractor's choice

It is for the contractor to choose to what extent he recovers his costs through his prices for 'preliminaries' and to what extent they are recovered from individual unit rates for items in other bills. For example, the cost of a concrete mixer could be recovered from an item for general provision of plant in the 'preliminaries bill' or from the rates for providing 1:2:4 or other mixes of concrete or mortar in the other bills. If he decides that it would be better to cover the item completely in the other bills, he should mark "INCL" against the item in Bill No.1 to indicate that he has 'included' for this expense elsewhere. There is no general right or wrong way for contractors to price preliminary items. The choice depends on their own methods of costing and estimating, the wording of the contract documents and their judgement as to whether billed quantities will be exceeded in the final measurement. If billed quantities are exceeded, they will gain financially by loading overheads onto the 'physical' bills. If the opposite applies, a full pricing of preliminary items is the better bet. A further factor is that some contracts allow the contractor to be paid for most preliminary items on early interim measurements. In such circumstances a bias to cost recovery through 'preliminaries' pays off in improved cash flow.

Items for Bill No.1

The items to be included in the preliminaries bill for our typical building are listed below. This list is neither exclusive or comprehensive, but is intended to indicate to the reader some of the main general items of expenditure to which a contractor is subject and which he may be given the opportunity to recover in this way. Other legitimate items of expenditure (such as his head office costs, salary and private car) will have to be recovered from percentage additions for overhead and profit to various bill items. A full wording for the items appears in the completed blank bills and summary sheet at the end of the chapter.

1. Plant, tools and vehicles
2. Site office
3. Storage shed °
4. Access and hardstandings
5. Fencing the site
6. Water supply
7. Safety, health and welfare
8. Employees' transport
9. Watchmen and other security measures
10. Scaffolding
11. Insurances
12. Performance bond
13. Clearing site and cleaning building.

Lump sums

It is not normally possible to put units and quantities against these general items. Thus they will all be marked LS so that the contractor can set a lump sum to cover all his expenses under the heading of the particular item.

Bills 1-7 and Summary

Now that we have 'figured out' the various items, we have a complete set of bills of quantities ready for the contractor to price. 'Putting a price on it' is the title of the next chapter, but the following few pages show how far we have got already in analysing the building design in a way that will lead to a scientific estimate of the cost of carrying out the work and a tender that will be both competitive and potentially profitable.

Item No.	Description	Unit	Quantity	Rate	Amount
	BILL No.1 PRELIMINARIES				
1.01	Allow for all necessary plant, tools and vehicles for carrying out the work	LS			
1.02	Provide temporary site office for the duration of the contract complete with table, chairs and all necessary office equipment and requisites.	LS			
1.03	Provide temporary shed for secure storage of goods, materials and components	LS			
1.04	Allow for preparation of access roads, tracks and hardstandings for the delivery, transport and storage of bulky materials	LS			
1.05	Provide temporary fence for the duration of the contract as supplied	LS			
1.06	Allow for providing a supply of water for the works	LS			
1.07	Allow for provision of all facilities, including latrines, to comply with statutory safety, health and welfare regulations in respect of all work-people employed (including employees of subcontractors)	LS			
1.08	Allow for transport of workpeople to the site and any other disbursements arising from their employment	LS			
1.09	Allow for expenses of watchmen and other security measures as may be deemed necessary to protect the works	LS			
1.10	Provide temporary scaffolding for the proper execution and completion of the works	LS			
1.11	Allow for obtaining 'Contractor's All Risks', public liability, employer's liability and other insurance cover in accordance with the terms and conditions of the contract	LS			
1.12	Allow for provision of contract performance bond to the value of 10 per cent of the contract sum in accordance with the terms and conditions of the contract	LS			
1.13	Allow for removing all rubbish and debris from the site and cleaning the building both internally and externally prior to handing over to the client	LS			

Transferred to summary sheet

Item No.	Description	Unit	Quantity	Rate	Amount
	BILL No.2 GROUNDWORKS				
2.01	Site clearance and removal of topsoil to approx. depth 10 cm	sq.m.	162		
2.02	Excavation to lower level of hard-core over building area, av. depth 10 cm	cu.m.	7		
2.03	Excavation in foundation trenches, av. depth 90 cm	cu.m.	22.5		
2.04	Prepare bottom of trench for concreting	sq.m.	12.5		
2.05	Return, fill and ram excavated material around foundation	cu.m.	17		
2.06	Place and compact hardcore 20 cm deep within building and blind with gravel to receive concrete	sq.m.	53.8		
2.07	Adjust levels, supply and roll gravel path 10 cm min. thick around building	sq.m.	52.5		
2.08	Excavation for access road, av. depth 30 cm	cu.m.	13.5		
2.09	Supply and compact hardcore to access road, min. 30 cm deep	cu.m.	13.5		
2.10	Supply and roll as specified gravel surface to access road 10 cm thick	sq.m.	45		
2.11	Level surplus soil as directed and tidy site on completion	LS			
	Transferred to summary sheet				
	BILL No.3 CONCRETE				
3.01	Formwork to sides of foundation base	sq.m.	12.5		
3.02	Formwork to sides of foundation walls	sq.m.	62.4		
3.03	Concrete 1:2:4 to foundation base	cu.m.	2.5		
3.04	Concrete 1:2:4 to foundation walls	cu.m.	6.2		
3.05	Concrete 1:2:4 to floor slab	cu.m.	4.1		
3.06	16 mm diameter mild steel rods in top of foundation walls including laps, bends and tying wire	kg	102		
3.07	8 mm diameter mild steel rods at 30 cm centres in both directions in floor slab including all tying wire, distance blocks and spacers	kg	142		
	Transferred to summary sheet				
	BILL No.4 BLOCKWORK AND PLASTERING				
4.01	Concrete block walls 20 cm thick in cement mortar	sq.m.	72.3		
4.02	Lintels over windows 20 x 25 x 140 cm reinforced with 4 No.6 mm diameter mild steel rods	No.	6		
4.03	Lintel over doorway 20 x 25 x				

Item No.	Description	Unit	Quantity	Rate	Amount
	240 cm reinforced with 4 No.6 mm diameter mild steel rods	No.	1		
4.04	External wall plastering 17 mm thick, to receive paint	sq.m.	77.0		
4.05	Internal wall plastering 17 mm thick, to receive paint	sq.m.	73.0		
4.06	Floor screed, average thickness 3 cm	sq.m.	53.8		
4.07	Supply, position and grout in place ½" anchor bolts for roof fixing	No.	30		

Transferred to summary sheet

BILL No.5 CARPENTRY AND WINDOWS

Item No.	Description	Unit	Quantity	Rate	Amount
5.01	Supply and fix 1.00 x 1.25 steel windows as specified	No.	6		
5.02	Timber 100 mm x 100 mm wall plates, including fixing to anchor bolts	lin. m.	35		
5.03	Supply and fix roof trusses in accordance with details supplied	No.	3		
5.04	No.16 gauge hoop iron straps 32 mm wide and 40 cm long including screw fixing to wall plates	No.	6		
5.05	Supply and fix 50 mm x 100 mm purlins	lin. m.	88		
5.06	Supply and fix 31 mm x 150 mm fascia board	lin. m.	36		
5.07	Double doors 2.00 x 2.25 m including frame, hinges, lock and fitting, all in accordance with details supplied	No.	1		

Transferred to summary sheet

BILL No.6 ROOFING

Item No.	Description	Unit	Quantity	Rate	Amount
6.01	Corrugated galvanised mild steel roof sheets fixed to timber with ¼" diameter galvanised roofing screws 2½" long each with one diamond-shaped bitumen washer and one galvanised steel washer. Side laps of 1½ corrugations	sq.m.	85.8		
6.02	Ridges 38 cm girth, including fixing as specified	lin. m.	11		

Transferred to summary sheet

BILL No.7 PAINTING AND GLAZING

Item No.	Description	Unit	Quantity	Rate	Amount
7.01	Glazing to windows as specified	No.	6		
7.02	Paint external surface of walls as specified	sq.m.	77		
7.03	Paint internal surface of walls as specified	sq.m.	73		

Item No.	Description	Unit	Quantity	Rate	Amount
7.04	Paint windows as specified	No.	6		
7.05	Paint doors and frame as specified	No.	1		
7.06	Paint fascia boards as specified	LS			

Transferred to summary sheet

SUMMARY SHEET

Bill
No. $

1. Preliminaries
2. Groundworks
3. Concrete
4. Blockwork and Plastering
5. Carpentry and Windows
6. Roofing
7. Painting and Glazing

Chapter Six

Putting a Price on it

The estimate

In the last chapter we started with the contract drawings for a small building, and used them to produce a bill of quantities for the work that will be involved in constructing it. This is the first stage of estimating — breaking down the work involved into individual tasks and quantities. Sometimes this first stage is done for the contractor, and a bill of quantities arrives on the contractor's desk as a part of the contract documents sent out (and prepared by) the client's professional representative. But the second stage of estimating — 'putting a price on it' — can only be done by the contractor himself. Only 'Contractor A' can forecast what it will cost *him* to get the necessary resources of men, materials and plant to the site and get them organised to complete the building on time. 'Contractor B' would be likely to come up with a rather different answer, as would Contractors C, D, E, F, etc.

Why estimates differ

Why do different contractors come up with different estimates? Even in a perfect world, where every contractor had all the resources he required and no contractor ever made a mistake, no two estimates would be exactly the same — because different contractors operate from different locations and have different ranges of skill and expertise. In the real world, where resources are always constrained by the availability (or lack!) of finance, and skills and efficiency vary widely, it is inevitable that unit prices should vary even more from contractor to contractor.

From estimate to tender

These differences are moderated to some extent in the final stage of the process, when the estimate is turned into a tender by making an appropriate addition for overheads and profit. The highly efficient contractor can allow a very comfortable percentage addition for profit to his basic estimated costs, happy in the knowledge that his overall price will still be competitive. The incompetent contractor, however, starts with the handicap of high basic costs. He will only stand a chance of getting the contract if he cuts his profit margin to the bone. This illustrates the discipline imposed on the businessman/contractor by the system of competitive tendering. Overall tenders have to be at about the same level if a group of competing contractors are all to get a reasonable share of the work that is available. They are all subject to the fundamental equation of contracting:

$$COSTS + PROFIT = TENDER \ SUM$$

So the only route to high profits and steady expansion is through increased efficiency and low costs. There is no doubt about it — for the contractor — EFFICIENCY PAYS.

Be realistic!

Efficiency is achieved on the site — not on paper. Only the operational and supervisory staff can cut operating costs, since only they are in a position to use the available resources to the best possible advantage. The job of the estimator is not to cut costs on paper, but to make the most *realistic* possible estimate of what it will cost his contracting firm — with its own peculiar mix of advantages and disadvantages — to carry out the work required. A low estimate is no compliment to the estimator's skills. It is dangerous for his firm, since it gives the staff a false sense of security and may lead to the

126

award of a contract that will end up showing a loss. 'Let's pretend' is a game for children – for grown-up contractors it is the first step on the road to bankruptcy.

The example

Returning to our example, we will try to produce a realistic tender for our own imaginary contracting firm. We will of course be making a lot of assumptions, and the reader is reminded that this is not an example to follow slavishly. Wage rates and material prices vary so much between countries and even between towns and areas in the same country that it just would not be possible to produce a set of universal prices. The second point is that this is a very simple, basic building – without even a water supply, drainage or an electricity service. The point of the example is to serve as a simple *introduction* to estimating and tendering, demonstrating the way in which the estimator learns to *think* and approach his task.

Learning by doing

The reader will not learn to be an estimator by just sitting in a chair and reading this book. The author's intention is much more modest and (he hopes) realistic. It is intended that this book should take the 'mystery and magic' out of estimating, and show that any contractor who knows his costs (and if he doesn't know his costs, he isn't a real contractor!) can sit down at his desk and produce a reasonably accurate estimate for the sort of work he is used to carrying out. The first time it may seem a bit difficult, but with care, persistence and attention, it will be finished. Maybe he will ask a more experienced friend to check it over before he sends it in to the client. The second time he will be a bit quicker and gain more confidence. This is the process of 'learning by doing', which is the only possible way of learning a practical skill. After a few months practice, the present example should appear ridiculously simple and the contractor will have a much better 'eye for costs' which will help him in his tasks of purchasing and site management as well as improving his tendering skills still further.

How to go about it?

Before we start to put a price on the job, we need to decide on how we are to go about organising and executing the work. This means that we need to produce a programme for the contract, at least in outline. The programme need not be too complicated, as it is to be used mainly as a guide to the duration of various activities for estimating purposes. A simple bar chart will generally suffice. However, contractors should note that it is becoming a more common practice for clients to require the tendering contractors to submit a draft programme with their tender, as an indication that they have given some realistic thought to how they would plan their operations if their tenders were to be accepted.

Contract period

In the example we shall assume that the client is allowing a maximum contract period of three months. This should allow comfortable time to complete such a simple building, and we will plan to complete in twelve weeks leaving one week spare to cope with any risks or contingencies that may arise. Some contract documents require the bidders to state their own period for completing the work. The client then chooses the most favourable bid on the combined grounds of price and time, and the lowest bid might be passed over in favour of an offer to complete the work more quickly. In such a situation the prospective bidder must give careful thought (based on careful planning) to his proposed contract period, since a failure to meet a self-imposed contract deadline would be bad for his reputation and could incur liquidated damages. Building is a risky business and, even though competition is keen, it always pays to allow a little 'slack time' in every programme to allow for unforseen contingencies. Thus, even on the inevitable occasional contracts where the worst comes to the worst, the work should be completed within the time stipulated. A reputation for 'completion on time' is as valuable an asset as money in the bank for the serious long-term contractor.

More about programming

A suggested progress chart for our example is shown on the following page. For more information on programming techniques, the reader should consult the companion volume *Financial Planning for the Small Building Contractor* (Intermediate Technology Publications Ltd., 1979).

Visit the site

Before he gets down to the job of working out costs, a serious contractor would first visit the site of the project to assess the factors that will influence his costings. No contractor should put his name to a tender without a clear understanding of the practical difficulties that will have to be faced in carrying out the work. Even the most foolish

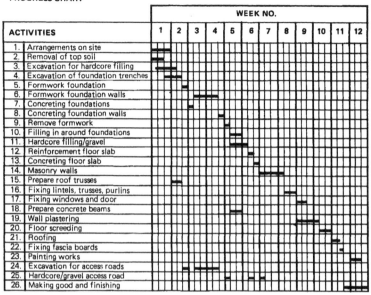

	ACTIVITIES	1	2	3	4	5	6	7	8	9	10	11	12
1.	Arrangements on site												
2.	Removal of top soil												
3.	Excavation for hardcore filling												
4.	Excavation of foundation trenches												
5.	Formwork foundation												
6.	Formwork foundation walls												
7.	Concreting foundations												
8.	Concreting foundation walls												
9.	Remove formwork												
10.	Filling in around foundations												
11.	Hardcore filling/gravel												
12.	Reinforcement floor slab												
13.	Concreting floor slab												
14.	Masonry walls												
15.	Prepare roof trusses												
16.	Fixing lintels, trusses, purlins												
17.	Fixing windows and door												
18.	Prepare concrete beams												
19.	Wall plastering												
20.	Floor screeding												
21.	Roofing												
22.	Fixing fascia boards												
23.	Painting works												
24.	Excavation for access roads												
25.	Hardcore/gravel access road												
26.	Making good and finishing												

contractor would not buy a second-hand truck without looking it over carefully, but some novice contractors are quite prepared to put much larger sums at risk in projects without taking this simple precaution.

Take a notebook — and the drawings

The first thing to check is that the drawings show everything that may affect the job. Some architects' 'standard' drawings just could not be used without amendment to suit the particular site, and the architect or engineer responsible should be asked to clarify them in such a case. The notebook is important because there will be a lot to see — and remember — and the estimate must be based on facts rather than memories.

Access

Examine the general access to the site. For example, the only access through the local village might be through a narrow archway. Note its size, so you can check later on whether your truck can get through it. If not, materials might have to be carried or brought in on barrows at much greater expense. Will the road be passable during the rainy season? Check also on possible obstructions on and around the site, such as trees and streams.

Ground conditions

It is important to know the type of soil and it's stability. If in doubt, dig a trial hole (after having obtained permission!). Can surplus soil from the excavation be easily and cheaply tipped nearby?

Services

Check whether clean water for concrete mixing, etc. is available nearby and where the nearest electricity supply is located should you need it to run your plant and equipment. Is there a local garage that can supply diesel fuel for plant and assist with emergency repairs?

Labour

If you do not know the area well, it would be worthwhile to find out whether reasonably hard-working casual labour can be recruited as an alternative to transporting in your own labour.

General site conditions

Take a good look at the site and work out in your mind how you will go about organising work on the site if you obtain the contract. How much plant and equipment will be required, and will it be necessary to buy or hire additional items to get the work done? What provision will have to be made for site security? Is theft or vandalism likely to be a problem? If any other thoughts occur to you about how to go about organising the job — write them down! They will be a big help later in ensuring that your estimate is an accurate reflection of likely costs.

Prices and costs

Our objective is to decide on realistic unit rates and lump sum prices to enter against all the various bill items that appear at the end of the previous chapter. To do this we need to know the real costs to the contractor of his resources of labour, materials and plant. Note that there will be a difference, sometimes a large one, between apparent cost and real cost. The apparent cost of a labourer is his daily wage, but his real cost must take account of additional costs of taxes, national insurance, holiday pay schemes, non-productive time and direct supervision costs. These additional costs vary from country to country according to national labour laws, the custom in the industry and any agreements that may have been negotiated with the trade unions. Equally the

The Site Visit.

1. TAKE A NOTE BOOK WITH A SKETCH PLAN OF THE BUILDING SITE.
2. HAS ANYTHING BEEN MISSED FROM DRAWINGS?
3. LOCAL SURROUNDINGS: TREES, STREAMS, OBSTRUCTIONS
4. SOIL TYPE AND STABILITY.
5. IS WATER FOR THE WORKS AVAILABLE NEARBY?
6. IS LOCAL LABOUR AVAILABLE?
7. GENERAL SITE CONDITIONS.

material costs that matter are not just the price paid in the shop or merchant's yard, but the cost on the site after allowing for transport, storage, handling and wastage. The cost of plant is not just the cost of fuel and operator's wages, but the full cost (including depreciation) *per productive hour worked.* Too many contractors fail to appreciate their real costs, with the result that they submit low bids that may not even recover their overheads let alone leave them with a real profit to reinvest in their business.

The difference between Apparent and **Real** costs

LABOUR. REAL COST.

MATERIALS. REAL COST.

PLANT.

'All-in' rates for labour

Almost every item in the bill of quantities involves physical work carried out by skilled and/or unskilled workers. Thus, before we start to look at individual items, it makes sense to work out general 'all-in' rates for both skilled and unskilled labour for use in pricing these items.

Two rates

To simplify the tendering process we will divide our labour force into just two broad categories — skilled and unskilled — and work out an 'all-in' rate for each. The skilled group will cover all craftsmen, including carpenters, masons and painters, and the unskilled group will cover general labourers. In real life a carpenter might command a slightly higher rate than a painter, but estimating always requires some degree of simplification and if we choose a reasonable average daily rate the 'plus' and 'minus' errors will more or less cancel each other out. An estimate is not the same as a thesis for the degree of doctor of philosophy, which has to be accurate in every respect. It is a simple working document, and part of the skill of the experienced estimator is in knowing when he can 'cut corners' in his calculations by taking averages and making assumptions.

What's 'all-in'?

The reason for describing the labour rate we are trying to calculate as 'all-in' is that it is meant to include all the direct and indirect payments, overheads and oncosts that are associated with employing a labour force. It is vital that none of these extra costs are forgotten in calculating the 'all-in' rate, as they can only be recovered bit by bit from the labour content allowed for in the rate set against each item in each bill.

Total labour content = Total labour cost

When all the bills have been completely priced (assuming the estimate has been correctly prepared), the total of all these labour contents should provide a sum of money equivalent to the 'all-in' cost to the contractor of the wages and extra costs of providing labour on the contract.

Four headings

We can classify labour costs under four main headings as follows:

— Direct payments to employees
— Enabling expenses

— Costs when not present
— Costs of non-productive time

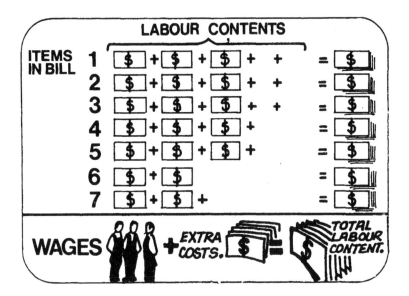

Direct payments to employees

The major direct payment to the employee is his wage or salary. But there may be additional payments made in certain circumstances, such as 'tool money' for craftsmen who provide their own tools, 'dirty money' for working in unusually unpleasant conditions (e.g. rodding out sewers) or 'danger money' to compensate for the employee undertaking work where there is some risk to life and limb. Where such payments are made, it is normally only to part of the workforce for part of the time and it can be allowed for in the 'all-in' rate as an average percentage addition to the standard daily wage. If such payments were on average 50 cents per day but were only expected to be paid on about one-fifth of the total number of working days, the expected cost could be recovered by an addition of $(\frac{50}{5})$ or 10 cents to the basic rate in calculating the all-in rate.

Overtime and guaranteed bonus

Additional overtime and guaranteed bonus payments also go direct to the employee and must be allowed for under the first heading. If overtime is only paid for at standard rates there is no need to make any additional allowance, because the overtime hours paid for will be recovered directly from the work done by the employee in those hours. But if

133

overtime is paid for at enhanced rates of 'time and a quarter', 'time and a half' or even 'double time', any work done on overtime will be a dead loss to the contractor if his estimate of labour cost was based on basic wage rates. It is not usually possible to foresee exactly the number of overtime hours that might have to be worked, except on 'emergency contracts' with very short contract periods and severe liquidated damages provisions. Thus in most cases the extra cost of possible overtime payments can be covered by a modest percentage allowance based on previous experience.

Bonus payments
Bonus payments linked to productivity do *not* require any special addition since they are 'paid for' by the employee himself who only qualifies for bonus if he finishes the work more quickly than allowed for in the estimate. But 'standing bonuses' which are not linked to output must, if they are likely to be paid for some reason, be allowed for in the 'all-in' rate.

Enabling expenses
The phrase 'enabling expenses' has been used to cover those cash costs of employing labour that do not end up in the employee's pay packet. They include government taxes and levies and (if paid) such items as travelling expenses to the site and lodging expenses. For estimating purposes they are again best dealt with by a realistic percentage addition to the average basic rate.

Cost per day paid
To simplify our example we will assume average basic wages as follows:

Skilled — $10 per day
Unskilled — $5 per day

We will also assume that the percentage addition to this basic wage to cover other direct payments and enabling expenses should be 20 per cent. Thus we can now work out the *cost per day paid* as follows:

	Unskilled	Skilled
Basic daily wage	$5.00	$10.00
Add 20 per cent	1.00	2.00
Cost per day paid	$6.00	$12.00

Cost per day worked
Unfortunately for the contractor, men are not robots and

they do not work all day and every day. Thus the cost of labour *per day paid* is by no means the same as the cost of labour *per day worked.* To get from one to the other we have to make allowances for the 'costs when not present' and the 'costs of non-productive time'.

Costs of non-productive time

We have to be realistic and recognise that our employees are human beings, so they will occasionally take off some time for 'tea breaks' and 'lunch breaks' and some of them may even just disappear off the site for a time if the foremen isn't looking. Bad weather can also cause hold-ups, and bad management can be yet another cause of non-productive waiting time if materials and/or instructions are not given when they are needed. Non-productive time varies a lot as it depends on the efficiency of the supervisory and management staff, but it is often far higher than the contractor believes it to be. For our example we will take a figure of 15 per cent as follows:

	Unskilled	Skilled
Cost per day paid	$6.00	$12.00
15% non-productive time	90	1.80
	$6.90	$13.80

Costs when not present

Depending on the laws of the country, the contractor may be required to go on paying his staff on national holidays even if they are not at work. There may also be a provision that he should grant a paid annual holiday to all permanent employees. Then there may also be a requirement that the employer should grant paid sick leave to any employee who can provide a doctor's certificate. Thus instead of working the 52 weeks a year for which he is paid, the average employee is likely to be present on the site for only, say, 48 weeks. Thus we need to increase our daily rate again to ensure that 48 weeks' work is enough to pay for 52 weeks' cost:

	Unskilled	Skilled
Cost brought forward	$6.90	$13.80
Multiply by $(\frac{52}{48})$ to give *all-in rate*	$7.47	$14.95
For convenience, round off to	$7.50	$15.00

Supervision

A further addition can be made to our calculated 'all-in

rate' to cover the cost of supervision. This can be done by spreading the 'all-in' costs of the site foreman and other supervisory staff over the remaining employees on the site (e.g. a site foreman with an 'all-in' daily cost of $22 supervising a site of 20 men could be paid for if *their* all-in costs went up by $(\frac{\$22}{20})$ or $1.10). But it is equally effective to recover all supervisory costs through general overheads, and that is what we will do in the present example.

The estimate
Now that we have decided on the daily cost to our contracting firm of each of our craftsmen and general labourers (which has turned out to be almost half as much again as the 'apparent cost' shown by their daily wages!), we can go on to estimate the rates to be set against the items in the bills of quantities. At first we will concentrate on estimating actual direct *costs.* Then we will decide on a percentage to be added to cover overheads and profit, which will give the *rates* to be entered in the bill of quantities and used to calculate the *bid price* we will submit to the client.

Bill No.1. Preliminaries
We will start at the beginning with Bill No.1. All the items in this bill are lump sums, and we will try to work out estimated costs for each item.

1.01 Plant, tools and vehicles
It is open to the estimator to decide whether to cover these costs under this item in Preliminaries or to include them as they are used under items in other bills. There will be little need for mechanical plant on a simple building contract of this kind, and the only item that is likely to be required is a concrete mixer. The cost of the mixer will naturally have to be covered somewhere. If it is hired, the cost will consist of hire charges, delivery charges and fuel. If the firm's own mixer is used it is best to work out an 'internal hire charge' to cover depreciation, maintenance, repairs, etc. and this will be the basis for charging clients on the various contracts and ensuring that the mixer 'earns its keep'. In this case we will cover the 'internal hire charge' and running cost in the concreting items, but use this item to cover the cost of setting up the mixer in a convenient place on the site. This estimate will be approximate (one stage up from a guess!), so we shall say:

Transport (contribution to cost since
only part of a full load)   say $ 5

Labour: 1 carpenter + 1 labourer		
for ½ day; i.e. ½ ($15 + $7.50)		$11.50
Timber base (railway sleepers)	say	$10
Dismantling and removal	say	$ 8
Total		$34.50
To round it off to the nearest dollar,	say	$35

Tools

There will be a need for a wide variety of hand tools and simple equipment to carry out the work such as wheelbarrows, shovels, picks, axes, sledgehammers, buckets, brushes, ropes, etc. If they are properly looked after these small items will last for years and go on being used on one site after another. Thus it would not make sense to write off their full cost on a single job. However, nothing lasts forever and these items represent a part of the firm's capital which will have to be repaired and replaced over a period. It would be a waste of time to set an arbitrary lifetime for every saw, chisel and hammer and calculate a few cents depreciation for each. Thus on the basis of experience we will put in a figure of $80 to cover losses, breakages, repairs and new purchases of small tools on this contract — and by making a similar modest charge on every contract we undertake enough funds will be generated over the course of the year to keep the general stock of tools in good condition.

Transport

As with plant costs, most transport costs can be included in the relevant items in other bills. For example, the materials costs of cement, sand and aggregate in concrete can be based on "as delivered" rather than "ex works" prices and an allowance can be made for the cost of handling and movement around the site. If the contractor uses his own car or pick-up to collect miscellaneous small items for his sites this cost will be reflected in general overheads. But on larger contracts, particularly road and civil engineering works, where a truck is allocated permanently to the contract, the cost of the vehicle must be recovered directly from that contract and specifically allowed for in the estimates.

Occasional use

The cost of making a vehicle permanently available could not be justified on the contract for the small building in our example. We will assume that the delivery of materials is to be recovered from the materials contents in the appropriate rates, and that any minor transport needs are covered by the use of the contractor's own pick-up and covered in our later percentage addition for general overheads. However we will assume that a three-ton tipper truck is hired to bring sheds, tools, etc. to the site and for any occasional transport needs that might arise.

Lorry hire
The basis for lorry or truck hire charges varies considerably from place to place. Sometimes there is an hourly or daily hire rate, sometimes the charge is based on a set rate per kilometre for the vehicle concerned and sometimes there is a mixture of the two or a rate per 'ton kilometre' in which you multiply the rate by weight *and* the distance to obtain the charge.

Transport estimate
The site is about 10 km from the contractor's home town, so every round trip will be 20 km. If the hire charge for a tipper is $2 per kilometre, the cost per round trip to the site will be $40. This is quite a lot of money, and is an indication of how important it is to pre-plan all transport movements and avoid unnecessary journeys. For estimating purposes we will assume three round trips from town to site, so this element in the estimate will be:

3 x $40 = $120

Estimate for Item 1.01
We can now complete the estimate for Item 1.01 as follows:

Plant	$ 35
Tools	$ 80
Transport	$120
Total Item 1.01	$235

1.02 Site office
The contractor is required to provide some form of site office in which to keep drawings, site records and instructions. The site office will also be used for discussions and meetings with the architect and clerk-of-works, and perhaps for negotiations with suppliers and sub-contractors. It is a false economy to try to get by without a site office of any kind. Not only does the absence of an office give the bad impression of an irresponsible 'fly-by-night' contractor, but the absence of place to keep, update and study drawings and records may result in serious errors in setting out or construction. By the time that these errors have been put right,

the cost to the contractor will be far more than he could possibly save by omitting to provide a reasonable temporary site office.

However, the office need not be lavish, particularly in this case where the site is so close to the contractor's head office, and a temporary wooden building of, say, 2 metres by 2 metres would be ample. Most established contractors own a number of these temporary timber buildings, and move them from site to site as the need arises.

We shall prepare a simple estimate for such a building but, since it will be used again and again on other sites in the future, it would not be fair to charge its complete cost against this one job. Thus we will divide the estimated cost of the building by the estimated number of uses to get an appropriate charge against the present contract. To this we will of course have to add the full cost of erecting and dismantling the building on this particular site.

We will assume that the building has a wooden frame consisting of 100 mm x 100mm and 100mm x 50 mm timbers, with timber plank walls and a corrugated iron roof. There will be one window in the side, and a lockable door will be provided at one end. The rough cost estimate is:

Timber	say,	$200
Window	say,	20
Door, with lock	say,	40
Corrugated iron	say,	50
Nails, screws, etc.	say,	15
Total		$325

If we assume that the building can be re-used on five future sites, this means that it will be depreciated over a total of six uses. Thus the element of the estimate to cover the building itself should be:

$$\frac{\$325}{6} = \$54.$$

Now we have to estimate the cost of preparation, erection and dismantling:

Level ground and prepare gravel base:
2 labourers for one day: 2 @ $7.50	=	$ 15

Erect office:
2 carpenters for one day: 2 @ $15	=	$ 30

Dismantling:
1 carpenter for ½ day: ½ @ $15	=	$ 7.50

Materials (gravel fill)	say	$ 7.50
Thus cost of preparation, erection and dismantling	=	$ 60
Add share of cost of building as above		$ 54
Sub-total		$114
Add allowance for furniture, shelves, office equipment and requisites (mostly re-usable)	say	$ 50
Total Item 1.02		$164

1.03 Storage shed

The cost of providing our 2 m by 2 m office building in the above item worked out to $114 for 4 sq. metres of building, or $28.5 per square metre. The storage shed will be simpler and less well-finished, so its cost should be within $20 per square metre. If we require a store of 2 m by 3 m, we can therefore get a reasonable estimate on a cost/unit area basis as:

$$2 \times 3 \times \$20 \qquad = \qquad \$120$$

1.04 Access roads, tracks and hardstandings

Since this is a simple building, it should not be necessary to provide an extensive network of temporary access roads on the site and it should be possible to save expense by making use of the permanent access roads for deliveries and parking of motor vehicles.

The positioning of hardstandings and storage areas for aggregates and other materials requires forethought to minimise the need for double-handling and unnecessary movements around the site. Double-handling of materials is a major hidden contributor to site costs, and the contractor — and his supervisors and other site staff — must understand that the real cost of materials rises every time they are moved. *Real costs* of materials are the costs when they are actually incorporated into the structure of the building, including all the costs of transport, handling and wastage. If they took the trouble to work it out, many contractors would be frightened to find that the *real costs* of many of their materials, like aggregates and concrete blocks, can be half as much again as the price they pay to the supplier.

Whilst there is no point in spending too much money on access and storage areas, it is worthwhile to make sure that the areas that are to be used for these purposes are properly prepared. For example, the area to be used for an aggregate stockpile should be free of topsoil and levelled to minimise wastage and the risk of contamination. It will also be worth-while to spend a little money on labour, hardcore and gravel to ensure that blocks, frames, timber, etc. do not get dirty or damaged before they are used. Thus under this item we will make an allowance as follows:

Labour: 2 general labourers for 2 days		
@ $7.50	=	$30
Materials: Hardcore/gravel	say	15
Estimate for Item 1.04		$45

1.05 Temporary fence

The contract does not call for the provision of a per-manent fence around the site, but it does require that a temporary fence be erected to protect the works during the construction period. We will assume that the fence is to be 1 metre high and 40 metres long with two strands of barbed wire for additional protection. The fence will be set on poles 2 metres apart, and a single gate 3 metres wide will be provided at the entrance to the main road. Since the fence is temporary, all the materials can be recovered and re-used on future contracts and we will therefore depreciate the fencing materials over 5 uses. Thus the calculation of estimated cost will be as follows:

Materials

Fencing	40 metres @ $5		$200
Poles	20 metres @ $1.90		38
Barbed wire	80 metres @ $0.50		40
Gate with posts		say	25
Sub-total			$303

Since we assume the fence will be used 5 times, our materials cost will be ($\frac{$303}{5}$)   say $ 61

Labour

Erection: say 2 labourers for 1½ days		
@ $7.50		$ 22.50
Dismantling: 2 labourers for ½ day		
@ $7.50		$ 7.50
Total Item 1.05		$ 91

1.06 Water supply

A supply of clean water is essential both to comply with the conditions of contract and to ensure that the concrete and mortar mixed on the site is of adequate quality. A piped water supply to this site with a standpipe is the best solution, but in remote areas it is sometimes necessary to fetch water by hand or even by lorry in barrels. If the quotation for a piped supply is too expensive, it is necessary to make a rough calculation of the volume of water that will be consumed in order to work out the number of trips and therefore the cost of bringing water to the site. In this case we will assume that a sufficient supply can be obtained by standpipe at an estimated cost of $180 including the installation and the estimated cost of water consumed.

1.07 Safety, health and welfare provisions

In some countries wide-ranging provisions are laid down by law and strictly enforced by Ministry of Labour inspectors who have the power to close a site down if they are not satisfied. In other areas there is a traditionally slack approach to safety and welfare, and contractors can get away with little or no expenditure under these headings. Such a miserly approach is short-sighted in the long-run, however, because workmen tend to perform better and show greater loyalty to their employer if they are treated as responsible human beings rather than mere tools of production. At least the contractor should provide a proper latrine, hand-washing facilities and a first aid box in case of accidents. We will allow a sum of $80 under this heading.

1.08 Transport of workpeople and other disbursements

We have included an allowance for travelling expenses in our 'all-in' labour rate and also covered other known or likely payments, so we can write *INCL.* against this item to show that these costs have been *included* in other items elsewhere in the bill of quantities.

A tactical reason for including these costs in the 'all-in' rate rather than pricing them separately is that some client's representatives require that evidence that money allocated under Preliminary items has actually been expended before they will authorise payment. Thus a contractor who has priced for bringing in his workmen by lorry from the nearest town but manages to recruit local people instead will lose out if he had priced this item. But if he had included the same amount of money in his 'all-in' labour rates, which in turn disappeared into the overall rates in various bill items, he would be able to take advantage of his saving because there would be no way in which it could be challenged!

1.09 Watchmen and other security measures

This is a difficult item to price. Even if watchmen are not

142

highly paid, they are non-productive and their wages add up to quite a lot of money over a 12-week period. Whether such expenditure is justified really depends on the value of the building and the materials and plant stored on the site, and also on the general reputation of the area. In city centres where theft and vandalism is very common, even watchmen may not be enough and it may be necessary to hire security men with guard dogs. In a peaceful village community where everyone knows everyone else, the risk will be very much less — particularly if the new building will benefit the local community by bringing new employment or welfare facilities. We will assume the site is in 'friendly territory' and, since we have already allowed for a lockable office and store and a secure fence around the site, it should not be too risky to leave it at that and write *INCL.* against this item too.

1.10 Scaffolding

The contractor is required to provide scaffolding during the period of the wall and roof construction. Some in-experienced contractors think that they can save money by not supplying scaffolding and their employees have to make do as best they can with planks of timber resting precariously on upturned oil drums. This is definitely a false economy. First there is the safety angle. Besides the human misery that results from an accident, the injured employee would have a good basis for a damages claim against his employer if there was a refusal to supply scaffolding (and there may be criminal charges under safety regulations as well!). Secondly, men cannot work quickly and well under difficult conditions, and the cost of erecting proper scaffolding is generally recouped directly by better output and elimination of the risk of having to return to make good faulty work.

Timber pole scaffolding can be quite satisfactory, providing it is put together by someone who really knows his job. But steel tubular scaffolding has a longer life and is easier to handle, even if it is more expensive in the first place. In the larger towns and cities it is often possible to hire scaffolding by the week from specialist companies, and it is also possible to engage specialist scaffolding sub-contractors to erect and dismantle scaffolding. We will assume that our contractor has his own stock of scaffolding and that his own men have been trained to erect it, so the estimate for this item will be based on an 'internal hire charge' for the use of the requisite tubes, boards and fittings together with the costs of transport, loading and unloading, erection, maintenance and dismantling.

Whilst this could be worked out separately for every contract, the experienced estimator will have calculated an 'all-in' scaffolding cost per square metre for buildings of various kinds (and checked its accuracy against actual costs). We will assume a figure of $2.50 per square metre and, since the wall area was calculated under Item 4.01 to be 72.3 sq. metres, we will allow for scaffolding:

72.3 x $2.50 = $181

1.11 Insurance

The various types of insurance cover that are available to the building contractor are discussed in Chapter Eight. Some types of cover are required by law, others may be a condition of contract (as in this case) and others are voluntary but are a prudent way of covering at least some of the risks and hazards to which a contracting business is subject. Insurance brokers are usually happy to advise on appropriate forms of cover and give quotations at the estimating stage, and we will assume that in this case the contractor is advised that the premiums that will be charged in order to obtain suitable cover will be about $150.

1.12 Performance Bond

The question of obtaining performance bonds is also discussed in Chapter Eight. We will assume that the contractor has consulted his bank or insurance firm about the availability of a bond, and has been advised to include a sum of $100 to cover the fee that would be charged for the provision of a bond if the bid proves successful.

1.13 Clearing site and cleaning building on completion

We will assume that the transport provision in Item 1.01 will cover the removal of rubbish and debris from the site. Thus we will simply allow for the provision of 2 general labourers for 3 days to deal with this work, i.e.

2 x 3 x $7.50 = $45

We have now completed the *estimating* stage for Bill No.1, and the figures are summarised below ready for conversion into *bid prices* after we have come to a decision on an appropriate percentage to be added to cover overheads and profit:

Bill No.1 — Estimate Summary

Item	Unit	Estimate ($)
1.01	LS	235
1.02	LS	164

1.03	LS	120
1.04	LS	45
1.05	LS	91
1.06	LS	180
1.07	LS	80
1.08	LS	INCL.
1.09	LS	INCL.
1.10	LS	181
1.11	LS	150
1.12	LS	100
1.13	LS	45

Bill No.2 Groundworks

2.01 Site clearance and topsoil removal (162 sq.m.)

At this stage we need to work out a unit rate, i.e. the cost *per square metre* of carrying out this work. The work will be done by hand, so we do not need to worry about plant or material costs on this item. If we do not have accurate output figures based on records for similar work on past contracts, it will be necessary to build up an estimate from the beginning. We start by imagining a single typical general labourer at work on this task, digging and moving soil at a steady rate, and deciding what sort of area he could complete in one hour. It is important that we work with an *average* figure. No doubt an olympic athlete hoping to gain a gold medal would complete the job in record time. But that would be no good for our estimate, because we won't be employing olympic athletes! We just need a typical figure for a normal sort of man working under reasonable supervision — and for the purposes of this example we will take a figure of two square metres per hour.

If the working day is eight hours, this means that the daily output should be 16 square metres — so the estimated cost per square metre will be one-sixteenth of the 'all-in' daily rate for a general labourer, i.e.

$(\frac{1}{16})$ x $7.50 = $0.47

2.02 Excavate av. depth 10 cm (7 cu.m.)

This item will also be costed on the basis of the time taken by an unskilled labourer to excavate a total volume of one cubic metre in shallow excavation. Since the average depth of excavation is 10 cm, one cubic metre will require excavation over an area of about 10 square metres. This work should go a little faster than the previous item, since no clearance is involved and no special care has to be taken in removing and stockpiling topsoil, so we will base our estimate on a daily output of about 20 square metres, i.e. 2 cu.m./day.

Since the all-in rate for a general labourer is $7.50 per day the rate per cubic metre is simply:

½ x $7.50 = $3.75

2.03 Excavate in trenches, av. depth 90 cm

It would not be worth bringing a mechanical excavator to the site for such a small job, so we will again base our rates on hand excavation. Digging in trenches is bound to be more time-consuming than shallow surface excavation, since the soil is more firmly compacted and working conditions are less easy. We will assume that a test pit has been dug and that there is no chance that rock might be encountered — so this will be pick and shovel work again. We will also assume that the trench will be stable without strutting or timbering, so no allowance has been made for this. Having examined the test pit, the estimator decides that an output of 1.5 cu.m./day would be reasonable. Thus the rate per cubic metre will be:

$(\frac{1}{15})$ x $7.50 = $5.00

It should be noted that, although we have been talking about the output of one man for estimating purposes, we have not made any judgement on the actual number of men that will be employed on this task. What we do know is that, since one man should be able to dig 1½ cu.m per day, the complete task of shifting 22.5 cu.m should represent 15 man days work. For estimating purposes, it does not matter if three men work for five days or a five-man team completes the task in three days. This decision will be made at the detailed programming stage, and may be modified as an operational decision on the site depending on labour availability and contract priorities.

2.04 Prepare trench bottom for concreting (12.5 sq.m)

This work has to be done carefully to ensure that there is no wastage of concrete, and we shall assume that the work is done by a mason and a general labourer working as a team. Together they should manage to cover three square metres per hour, which would be equivalent to 24 sq.m per day. Thus the rate per square metre is:

$(\frac{1}{24})$ x ($15 + $7.50) = $0.94

2.05 Return, fill and ram (17 cu.m)

This is an item which is often skimped as workmen heap the excavated soil back around the foundation and neglect to ram the backfill in layers as specified. They are *not* saving time and money for their employer in the long run because, taking this building for example, the path outside the building will probably subside requiring additional gravel to be placed and compacted and — more seriously — the concrete floor may crack and require expensive remedial measures.

Adequate levels of compaction can *only* be achieved by filling in thin layers and thoroughly ramming *each* layer. Filling the whole trench and then applying the rammer only compacts the top crust, leaving the underlying layers to sink gradually over a period so that that top crust eventually sinks into the void that has been created. Thus we will not cut corners on the estimate for this item, and allow an hour per cubic metre of backfill for the team of mason plus labourer. So the daily output will be 8 cu.m, and the estimate per cu.m will be:

$$(\frac{1}{8}) \times (\$15 + \$7.50) = \$2.81$$

2.06 20 cm hardcore with gravel blinding (53.8 sq.m)

The price of hardcore tends to fluctuate quite rapidly depending on availability, but we will base our estimate on a price of $5 per ton delivered to site. 1 cubic metre of hardcore weighs about 1.65 tons, so the cost per cubic metre will be:

$$1.65 \times \$5 = \$8.25$$

We need a rate per square metre for hardcore compacted to a thickness of 20 cm. But we will not of course be able to buy hardcore 'ready-compacted', so it will be necessary to buy more than one cubic metre of hardcore if we want to end up with a compacted volume of exactly one cubic metre. Naturally hardcore is a variable material, but on average if we start with a delivery of 5 cubic metres of loose hardcore we will end up with a compacted volume of 4 cubic metres. This effect is illustrated overleaf.

This means that, if we want to end up with 4 cubic metres of compacted material it will be necessary to buy an extra 1 cubic metre to make up for the amount lost on compaction. So it is necessary to add an allowance of 25 per cent to our requirements, which can be done by basing our calculations

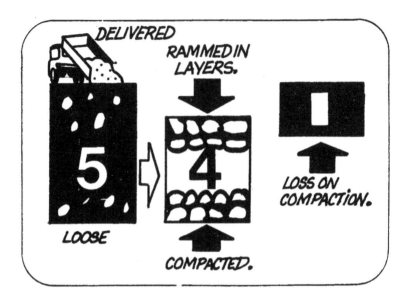

DELIVERED
RAMMED IN LAYERS.
5 LOOSE
4 COMPACTED.
LOSS ON COMPACTION.

on a loose thickness of:

20 cm x $(\frac{125}{100})$ = 25 cm

Thus a square metre will contain (1 x 1 x 0.25) = 0.25 cubic metres of loose hardcore. To that figure we will add a further allowance of 10 per cent to cover losses and wastage in handling, giving a thickness of

0.25 cu.m x $(\frac{110}{100})$ = 0.28 cu.m say

The other materials item is gravel for blinding and, including an allowance for compaction and wastage, we will calculate on the basis of a thickness of 10 cm (i.e. volume 0.10 cu.m) and a delivered cost per cubic metre of $8.00.

The calculation for a rate per square metre is:

Materials

Hardcore:	0.28 x $8.25	=	2.31
Gravel:	0.10 x $8.00	=	0.80
Sub-total			3.11

Labour
Barrowing, filling and ramming
(say ½ sq.m/man hour = 4 sq.m/man day)

	¼ x $7.50	=	1.88
Total			$4.99 per sq.m

2.07 Gravel path 10 m min. depth (52.5 sq.m.)

In addition to losses through compaction and wastage, there may be some low spots around the building which will require additional gravel fill. Thus we will base our calculations on a loose thickness of 16 cm, i.e. 0.16 cu.m/sq.m.

Materials
Gravel: 0.16 x $8.00   =   1.28
Labour
Barrowing, placing and rolling
(say 6 sq.m/man day)
                     ($\frac{1}{6}$) x $7.50   =   1.25
Total $2.53
                                                        per
                                                        sq.m

2.08 Excavation for access road, av. depth 30 cm (13.5 cu.m)

Under Item 2.02 we decided on a rate of $3.75/cu.m for excavation to an average depth of 10 cm. The rate for this item should be a little higher, though certainly less than for excavation in trench to an average depth of 90 cm (Item 2.03 — $5.00/cu.m), and a reasonable figure would seem to be $3.90/cu.m. There is no point in wasting time going over the same calculations time and time again, and the estimating process can often be speeded up by making a small adjustment to comparable unit rates that have already been calculated as we have done for this item.

2.09 Hardcore to access road, av. depth 30 cm (13.5 cu.m)

This item is priced *per cubic metre,* but we must remember that it will be measured in the compacted state so allowance will have to be made for wastage and compaction losses.

Volume when compacted   1.00 cu m
Add 25% for losses in compaction: .25
                                                1.25
Add 10% allowance for wastage   .13
Sub-total   1.38 cu.m
Materials
Hardcore: 1.38 x $8.25   =   11.39
Labour
Say 3 sq.m/man day. Av. depth
30 cm, thus 0.90 cu.m/man day)
                   0.90 x $7.50   =   6.75
Total $18.14 per cu.m

2.10 Gravel surface to access road 10 cm thick (45 sq.m)

Allow 15 cm nominal thickness to cover compaction losses and wastage, i.e. 0.15 cu.m/sq.m.

Materials
Gravel:	0.15 x $8.00	=	1.20

Labour
(Say 8 sq.m/man day)

$(\frac{1}{8})$ x $7.50	=	0.94
Total		$2.14 per sq.m

2.11 Level surplus soil and tidy site on completion (LS)

This item refers only to levelling and tidying excavated soil, since general site clearance is covered under item 1.13. There is no provision for transporting surplus excavated material off the site, although it may be possible to sell the surplus topsoil by arrangement with the client if there is a local demand for this material. We will price for 2 general labourers for 2 days, i.e.

2 x 2 x $7.50	=	$30

The estimated unit rates for Bill No.2 are now complete and are summarised in the following table:

Bill No.2 — Estimate Summary

Item	Unit	Estimate ($)
2.01	sq.m	0.47
2.02	cu.m	3.75
2.03	cu.m	5.00
2.04	sq.m	0.94
2.05	cu.m	2.81
2.06	sq.m	4.99
2.07	sq.m	2.53
2.08	cu.m	3.90
2.09	cu.m	18.14
2.10	sq.m	2.14
2.11	LS	30.00

Bill No.3 Concrete

3.01 Formwork to sides of foundation base (12.5 sq.m)

We will use timber formwork as shuttering for the foundation base and walls. In order to work out the quantity and price of timber required, it is best to sketch out how the carpenter will fix the formwork as shown opposite.

Materials

The formwork for the base is made up from 25 mm x 100 mm timber, with two planks laid one on top of the

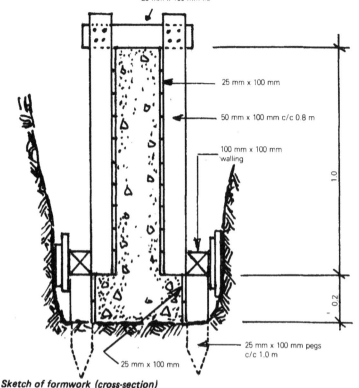

25 mm x 100 mm tie

25 mm x 100 mm

50 mm x 100 mm c/c 0.8 m

100 mm x 100 mm walling

1.0

0.2

25 mm x 100 mm pegs c/c 1.0 m

25 mm x 100 mm

Sketch of formwork (cross-section)

other running right round the base both inside and out and held firmly by pegs hammered into the base of the trench. To simplify the calculations we will not bother with precise lengths, but work with the building dimensions of 10 m x 6 m. Thus the distance around the perimeter of the building is:

2 x (10 + 6)   = 32 metres.

There are two planks inside and two outside so to get the length of formwork we multiply by 2 x 2 as follows:

2 x 2 x 32   = 128 metres.

Next we have to allow for the pegs. They are at 1 metre centres, so the number will be:

$(\frac{32 \times 2}{1})$ = 64 No.

We can now calculate the total length of 25 mm x 100 mm timber required:

Formwork: as above		128 m
Pegs: 64 No. x 0.40	=	26 m say
Sub-total		154

| Allow 10% for cutting and wastage | 16 |
| Total | 170 m |

Having obtained quotations from local timber merchants, we decide to base the estimate on a price of $0.90 per linear metre for 25 mm x 100 mm timber delivered to the site. Thus the cost of timber will be:

170 x $0.90   = $153

However it should be possible to use the timber more than once providing the formwork is dismantled and cleaned carefully. We decide that part will be re-used twice and part will be usable on three occasions, so we assume an average re-use of 2½ times and the costing against this contract is:

$\left(\dfrac{\$153}{2\frac{1}{2}}\right)$ = $61

| *Add* allowance for nails, etc. | 2 |
| Materials cost total Item 3.01 | $63 |

Labour
Allow one day's work for one carpenter and two general labourers:

1 x ($15.00 + 2 x $7.50)   = $30

| Add materials as above | 63 |
| Total | $93 |

Thus rate for Item 3.01 = $$\left(\dfrac{93}{125}\right)$$ = $7.44

3.02 Formwork to sides of foundation walls (62.4 sq.m)

Materials
This formwork and its associated supports is made up from 25 mm x 100 mm, 50 mm x 100 mm and 100 mm x 100 mm timber. The lengths of timber required are as follows:

25 mm x 100 mm
Formwork: Area 62.4 sq.m, so divide by width
to obtain length $\left(\dfrac{62.4}{0.1}\right)$ =   624 m

Ties: No. required = $\left(\dfrac{32}{0.6}\right)$ = 54

Thus 54 x 0.5	=	27 m
Wedges:	say	20 m
Sub-total		671 m
Allow 10% for cutting and wastage		
	say	69
Total		740 m

50 mm x 100 mm

2 x No. of ties = 108 No. x 1.20 m

=	130 m
Allow 10% for cutting and wastage	
say	15
Total	145 m

100 mm x 100 mm

Inside and outside, so: 2 x 32	=	64 m
Allow 10% for cutting and wastage		
	say	6
Total		70 m

The prices for 50 mm x 100 mm and 100 mm x 100 mm timber are $1.70 and $3.20 respectively, so the materials estimates (again assuming an average 2½ uses for the timber) will be:

25 mm x 100 mm:	740m @ $0.90	=	$666
50 mm x 100 mm:	145 @ $1.70	=	246
100 mm x 100 mm:	80 @ $3.20	=	256
Sub-total			$1168

Allow for 2½ times re-use ($\frac{1168}{2½}$)   = $ 467

Add for nails, etc.	say	13
Total		$480

Labour

Allow two days' work for two carpenters and three general labourers:

2 x (2 x $15.00) + (3 x $7.50)

= 2 x $52.50	=	$105
Total for 62.4 sq.m	=	$585

Thus rate for Item 3.02 = $($\frac{585}{624}$)   = $9.38

3.03, 3.04 and 3.05 Concrete 1:2:4 (12.8 cu.m total)

The first problem we meet is that we are required to price concrete on the basis of unit price per cubic metre, but the local merchants sell the cement, sand and aggregate that will be needed by weight. (Water will also of course cost the contractor something, but has already been covered in Item 1.06). The prices quoted (*not* including delivery) are:

Cement:	$70 per tonne
Sand:	$ 6 per tonne
Aggregate:	$ 8 per tonne

In order to carry out our estimate we need to know the conversion factors of weight to dry volume of these materials. These factors vary somewhat according to local conditions, but for the present example we will assume the following figures:

Cement:	1.4 tonnes = 1 m^3
Sand:	1.5 tonnes = 1 m^3
Aggregate:	1.6 tonnes = 1 m^3

On large contracts concrete is sometimes specified only in terms of required strength, and the contractor is required to design a mix that fulfills the requirements. In this case, however, a mix of cement, sand and aggregate in the proportions 1:2:4 is specified. We will calculate the cost of materials in a cubic metre of concrete in the same way as in the last chapter (although in this case the assumed prices are different). We will again assume 20 per cent shrinkage and ignore the volume of cement in the finished mix:

1 m^3 cement, i.e. 1.4 tonnes @ $70	=	$98.00
2 m^3 sand, i.e. 2 x 1.5 tonnes @ $6	=	18.00
4 m^3 aggregate, i.e. 4 x 1.6 tonnes @ $8	=	51.20
Sub-total		167.20
Add 25% to allow for 20% shrinkage		41.80
Total cost of materials for 6 m^3 concrete	=	$209.00

Thus cost of materials per cubic metre = $(\frac{209.00}{6})$ = $34.83

Transport of materials to site

Since the prices quoted for concreting materials are 'ex-works' rather than 'delivered', we will have to make an additional allowance in our unit price to cover transport costs. These will be based on a charge of $0.80 per tonne per kilometre. The distances from the various suppliers' yards to the site are as follows:

Supplier A (cement)	— 30 km
Supplier B (sand)	— 10 km
Supplier C (aggregate)	— 15 km

In order to calculate the 'transport cost content' in a cubic metre of concrete, we must first calculate the weights of each of the constituents. Using the figures set out in the materials cost calculation these are as follows:

Cement: $\frac{1}{6}$ x 1 x 1.4 x $(\frac{125}{100})$ = 0.292 tonnes

Sand: $\frac{1}{6} \times 2 \times 1.5 \times (\frac{125}{100})$ = 0.625 tonnes

Aggregate: $\frac{1}{6} \times 4 \times 1.6 \times (\frac{125}{100})$ = 1.333 tonnes

The transport cost component is now obtained by multiplying weight by the distance by the transport charge as follows:

Cement: 0.292 x 30 x $0.80	=	$ 7.01
Sand: 0.625 x 10 x $0.80	=	$ 5.00
Aggregate: 1.333 x 15 x $0.80	=	$16.00
Thus total cost of transport per cubic metre	=	$28.01

Plant

We assume that the specification calls for concrete to be mixed mechanically to ensure a more uniform mix. Thus the contractor will have to decide what size mixer to employ on the contract. Mixers are still generally described as 5/3½, 7/5, 10/7 etc. These two figures refer to the volumes *in cubic feet* of dry materials put into the mixer (the first figure) and mixed concrete yielded per batch (the second figure). The metric equivalents and outputs per hour (assuming a mixing cycle time of ten minutes) are given in the following table:

Mixer rating Input/output (cu. ft.)	Metric equivalent (m³)	Output per hour (m³)
5/3½	0.14/0.10	0.60
7/5	0.20/0.14	0.84
10/7	0.28/0.20	1.20
14/10	0.40/0.28	1.68

The smaller mixers are used mainly for mixing mortar (for which speedier cycle times can be achieved because there are only two ingredients) and the larger ones for mixing concrete on large contracts. For this contract a 7/5 mixer should be sufficient (the smaller the cheaper!) and we will assume that the specification permits batching by volume rather than by weight. The contractor's plant department quotes an internal hire rate for a 7/5 mixer of $6.00 per day plus delivery and collection charges of $20.00 each way.

With such heavy delivery and collection charges, it would not be worth sending the mixer off the site after each section of concreting is complete. The combined cost of collection and delivery is $40, so we would have to be sure that the mixer would be required for 7 full days (7 @ $6 = $42) for this to pay off. Thus we will estimate on the basis of the mixer remaining on the site for a period of 10 working days during which we will complete the concreting of the foundation and floor slab. Thus the cost of the mixer will be:

Delivery charge	$ 20
Hire charge: 10 days @ $6.00 =	$ 60
Collection charge	$ 20
Total	$100

The total volume of concrete in items 3.03, 3.04 and 3.05 is 12.8 cu.m, so the cost per cu.m is:

$\left(\frac{\$100}{12.8}\right)$	=	$7.81
Add for fuel and lubricants say		1.00
Thus cost of plant per cubic metre =		$8.81

We will assume that the mixed concrete is to be transported in barrows, since the cost of a motorised dumper would not be justified. Also no allowance will be made for mechanical vibration of the concrete as it will be placed and compacted by hand.

Labour

If it is to attain its maximum possible strength, concrete must be placed as soon as possible after it has been mixed. Thus the mixing point should be set up as close as possible to the placing area, and it is best to be generous with the allocation of manpower as the physical labour involved is strenuous and the materials are expensive. We will allocate a team of two skilled and eight unskilled men to this work as follows:

	Skilled	*Unskilled*
Loading mixer	—	4
Operating mixer	1	—
Barrowing concrete	—	3
Placing concrete	1	1
Total	2	8

The cost per day for this team will be:

2 unskilled at $15.00	=	$30.00
8 unskilled at $7.50	=	$60.00
Total		$90.00

during an 8 hour day would be 6.7 cubic metres. But there

156

are always occasional delays in practice, and it will be necessary to leave sufficient time to clean the mixer and tools thoroughly at the end of the day's work. Thus we will base the estimate on an average output of 4 cubic metres per day, so

Cost of labour per cubic metre = $\frac{\$90}{4}$ = \$22.50

We can now add the element of the estimate to obtain a rate for concrete per cubic metre:

Materials	\$34.83
Transport to site	28.01
Plant	8.81
Labour	22.50
Rate for Items 3.03, 3.04 and 3.05	\$94.15

The reader should note that this is in no way intended to be a typical price for concrete, since the cost elements vary enormously from place to place. It is intended, however, to show the reader a way to analyse the costs that he faces in his own operations and relate them to an estimate of a unit price that will allow them to be fully recovered.

3.06 16 mm mild steel rods in top of foundation walls (102 kg)

We will assume that the local steel stockholder quotes a price of \$700 per tonne for bars cut to length and delivered to the site. We will allow ½ day each for two skilled men unloading, bending, assembling and placing these bars.

Materials		
0.102 @ \$700	=	\$71.40
Add 5% allowance for wastage, ties, etc.		3.60
Sub-total		75.00
Labour		15.00
Add 2 x ½ x \$15		15.00
Total		90.00
Thus rate/kg = $(\frac{\$90.00}{102})$	=	\$0.88

3.07 8 mm mild steel rods in floor slab (142 kg)

We will assume a price of \$800 per tonne delivered for the rods, and allow a full day for two skilled men plus one general labourer for unloading and steelfixing.

Materials		
0.142 @ \$800	=	\$113.60
Add 5% allowance for wastage, ties, etc.		5.70
Sub-total		119.30

Labour

Add 1 x (2 x $15) + $7.50)   =   37.50

Total $156.80

Thus rate/kg = $\frac{$156.80}{142}$ = $1.10

We have now completed estimating the rates for Bill No.3 as follows:

Bill No.3 — Estimate Summary

Item	Unit	Estimate ($)
3.01	sq.m	7.44
3.02	sq.m	9.38
3.03	cu.m	94.15
3.04	cu.m	94.15
3.05	cu.m	94.15
3.06	kg	0.88
3.07	kg	1.10

Bill No.4 Blockwork and Plastering

4.01 Concrete block walls 20 cm thick (72.3 sq.m)

The walls are to be constructed of 25cm x 25cm x 20 cm thick cement/sand blocks set in 1:5 cement/sand mortar with 2cm joints as illustrated below:

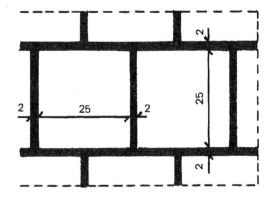

Materials

Each block (with its share of mortar surround) covers an area 27cm x 27cm. Thus the number of blocks per square metre is:

$(\frac{1}{0.27 \times 0.27})$ 13.72

Add 10% allowance for cutting and wastage   1.37

Total   15.09

158

We will assume that the contractor decides to buy in the blocks and obtains a quotation of $1.50 delivered to the site. Thus the cost of blocks per square metre is:

15.09 @ $1.50   = $22.64

Now we need to add the cost of the mortar. This is a comparatively low cost item, and could be dealt with by adding an estimated figure per square metre (say $2). However, in this example we will go through the full calculation, using the same method as was employed in the last bill to calculate the price of concrete.

For 6 cu.m of mortar:

1 m^3 cement: i.e 1 x 1.4 tonnes @ $70	= $ 98.00
5 m^3 sand: i.e. 5 x 1.5 tonnes @ $6	= $ 45.00
Sub-total	143.00
Add 25% to allow for 20% shrinkage	35.75
Total	$178.75

Thus cost/m^3 of basic materials = $29.79

Add for transport of cement and sand to site on same basis as for concrete ingredients:

Weight/cu.m:

Cement: $\frac{1}{6}$ x 1 x 1.4 x $\frac{125}{100}$ = 0.292 tonnes

Sand: $\frac{1}{6}$ x 5 x 1.5 x $\frac{125}{100}$ = 1.563 tonnes

Transport costs:

Cement: 0.292 x 30 x $0.80	= $ 7.01
Sand: 1.563 x 10 x $0.80	= 12.50
Total	19.51
Add cost of basic materials as above	29.79
Thus cost of mortar (materials)/cu.m	$49.30

To calculate the volume of mortar per square metre of finished wall we will start by calculating the 'share' of mortar relating to each block. This is:

0.01 x 0.20 x (0.27 x 4)   = 0.00216 cu.m

There are 13.72 blocks per square metre, so the volume of mortar per square metre must be:

13.72 x 0.00216	= 0.0296 cu.m
Add 15% for wastage	0.0044
Sub-total	0.034 cu.m

Thus cost of mortar per cubic metre = 0.034 x $49.30
= $ 1.68
Add cost of blocks as above   22.64
Total materials cost $24.32

Labour

We will assume that a team of two masons plus five general labourers can complete 8 sq. metres of wall per day, allowing for hand-mixing the mortar. If the contractor has a mixer free at the time, he may of course choose to bring it to the site to speed up the work and make a saving on labour costs. This will not affect the validity of the estimate, as by that stage the estimate should be seen as a target rather than a prescribed course of action and any savings that can be achieved are a sign of good opportunistic site management. At the present estimating stage we have to work with probabilities, and the cost per day of our proposed seven man team is:

(2 x $15.00) + (5 x $7.50) = $67.50
Thus the cost per square metre = $(\frac{\$67.50}{8})$ = $8.44

Add materials cost as above   24.32
Total Item 4.01 $32.76

4.02 Lintels over windows 20 x 25 x 140 cm (6 No.)

These are comparatively inexpensive items, and it should be possible to fit in the precasting of the lintels at the same time as the main concrete work to the foundations. For the concrete we will use the already calculated rate of $94.15/cu.m, which includes labour. For formwork a reasonable comparative rate would be $8.00/sq.m, also including labour.

Thus the estimated cost per lintel will be:

Concrete: 0.20 x 0.25 x 1.40 x $94.15
    = 0.07 cu.m x $94.15   = $ 6.60
Formwork:
    0.25 x 2 x (1.40 + 0.20) x $ 8.00
    = 0.80 sq.m x $8.00   = $ 6.40
Steel reinforcement, including fixing
say $ 3.00
Thus unit rate Item 4.02 $16.00

The labour contents in the above rates should be sufficient to cover the cost of lifting and building in the lintels.

4.03 Lintel over doorway 20 x 25 x 240 cm (1 No.)

There is only one unit of this item, so it is not worth

analysing in detail. Its cross section is the same as the previous lintels, so we will assume that the unit cost increases in proportion to the length, i.e.

$$(\tfrac{240}{140}) \times \$16.00 \qquad\qquad = \$27 \text{ say}$$

4.04 External wall plastering 17 mm thick (77.0 sq.m)

Plaster consists of a mixture of sand with cement and/or lime and the mixture depends on the price and availability of materials as well as the required strength and durability. In this case we will assume that the specification calls for a 1:5 cement/sand mixture (as for the mortar). Thus we can use the materials cost calculated under Item 4.01 as $49.30 per cubic metre.

We will base our calculations on a plaster thickness of 17 mm as specified, but increase the allowance for wastage, etc. to 20 per cent to allow for occasional areas where the thickness must be increased to keep the surface even.

Materials
Volume per sq.m: 0.017 cu.m
Thus cost per sq.m: 0.017 x $49.30   = $0.838
Add 20% allowance for wastage, etc   0.138
                                    $0.98 say

Labour
From records of similar plastering work on other sites, the contractor finds that the average output for one craftsman plus one labourer is 12 square metres per day, including preparation, mixing, etc. Thus the labour cost per square metre will be:

$\tfrac{1}{12} \times \$ (15.00 + 7.50)$ = $1.88
Thus unit rate for Item 4.04   = $2.86

4.05 Internal wall plastering 17 mm thick (73.0 sq.m)

The rate for internal plastering will be almost the same as for external plastering. The only difference is that we had allowed in Bill No.1 for scaffolding around the building, but we have so far made no allowance for internal scaffolding or trestles. Since the scaffolding will already be on the site, it will be sufficient to make a small allowance for the labour cost of erecting and dismantling scaffolding as required internally — say 6 hours (¾ day) for one skilled and one unskilled man, i.e.

¾ x $ (15.00 + 7.50)   = $16.88

This cost must be spread over the full quantity of 73.0 sq.m, so the full rate for this item is:

$2.86 + ($\frac{\$16.88}{73}$)   = $3.09

4.06 Floor screed, av. thickness 3 cm (53.8 sq.m)

For this item we again assume a 1:5 cement/sand mixture, so the materials cost of $49.30 applies here also.

Materials
Volume per sq.m: 0.03 cu.m
Thus cost per sq.m: 0.03 x $49.30   = $1.479
Add 15% allowance for wastage, etc.   0.224
                                                          $1.70 say

Labour
For this item we will estimate the time required for a given labour force to complete the job, and then work back to a unit labour rate. Assuming that no concrete mixer is available, we allow for hand mixing and assume a total of two days' work for a team consisting of two masons and four general labourers. Thus the total labour cost will be:

2 x $ (2 x 15.00) + (4 x 7.50) = $120.00

Thus the labour cost per square metre will be:
($\frac{\$120.00}{53.8}$)   = $2.23

Thus unit rate for Item 4.06   = $3.93

4.07 ½" anchor bolts for roof fixing

Materials
The materials cost for this item will consist of the price of the bolts and the cost of the grout. We will assume that the materials merchant quotes a price of $1.30 per bolt, and we will add 20 cents as a contribution to delivery and handling costs — making a basic $1.50 per bolt. It is not worth going to the trouble of calculating quantities and prices of grout, since only a little is involved. $7.50 should cover the total cost of the grout for this work, which is equivalent to 25 cents per fixing — bringing the unit materials cost to $1.75.

Labour
We will take the labour input as 1½ days for 1 mason and 2 labourers, so the unit labour cost will be:

$\frac{1}{30}$ x 1½ x $15.00 + (2 x 7.50)   = $1.50

Thus the cost for this item is:

$1.75 + $1.50 = $3.25

We can now summarise the unit rates for Bill No.4 as follows:

Bill No.4 — Estimate Summary

Item	Unit	Estimate ($)
4.01	sq.m	32.76
4.02	No.	16.00
4.03	No.	27.00
4.04	sq.m	2.86
4.05	sq.m	3.09
4.06	sq.m	3.93
4.07	No.	3.25

Bill No.5 Carpentry and Windows

5.01 Supply and fix 1.00 x 1.25 steel windows (6 No.)

The main cost element in this item will be the windows themselves. Quotations will be obtained at the estimating stage, and we will assume that the most favourable quotation for windows, including hinges and ironmongery, is $40 per unit delivered to the site. To this figure we need to add an allowance for handling, storage, lifting into position, setting and fixing. An allowance of one hour each for a skilled and an unskilled man should be sufficient, making:

$\frac{1}{8}$ × $(15.00 + 7.50) = $ 2.81

Add quotation for supply as above   40.00

Thus unit rate for Item 5.01 = $42.81

5.02 100 mm x 100 mm wall plates (35 lin.m.)

Materials

The quoted price for 100 mm x 100 mm timber is $3.20 per metre, and we will allow a 10% addition to the measured length for cutting and wastage:

$\frac{110}{100}$ × $3.20 = $3.52

Labour

We will suppose that the complete job will be done in one day by two carpenters assisted by one labourer. Thus the unit labour content will be:

$\frac{1}{35}$ × 1 × $ [(2 × 15.00) × 7.50] = $1.07

Add unit materials content as above   3.52

Thus unit rate for Item 5.02 = $4.59

5.03 Supply and fix roof trusses (3 No.)

Materials

Calculating the lengths of timber required in each of the trusses may seem complicated, but it is not too difficult provided the calculations are set out methodically. We will start with a sketch to identify the various members that go to make up the truss, and then calculate the lengths of timber required both for the members and the jointing plates that connect them. Although the lengths of all these individual members could be calculated, it is much quicker (and sufficiently accurate) to measure them off the scale drawing.

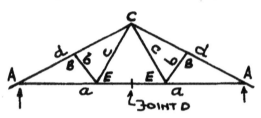

Main members:	Jointing plates:
a: 3.20 m	B: 0.30 m
b: 0.90 m	C: 0.50 m
c: 2.00 m	D: 0.50 m
d: 3.90 m	(No plates required for joints A or E)

The numbers and types of timber can also be taken off the drawing and classified as follows:

Main members:	Jointing plates:
a: 4 No.38 mm x 100 mm	B: 4 No.38 mm x 100 mm
b: 2 No.50 mm x 100 mm	C: 2 No. 38 mm x 200 mm
c: 2 No.50 mm x 100 mm	D: 2 No.38 mm x 100 mm
d: 2 No.50 mm x 100 mm	

We can now calculate the lengths of the three different sizes of timber, adding 15 per cent to the lengths shown above to allow for cutting and wastage:

38 mm x 100 mm	50 mm x 100 mm	38 mm x 200 mm
a: 4 x 3.20 = 12.8	b: 2 x 0.90 = 1.8	C: 2 x 0.50 = 1.0
B: 4 x 0.30 = 1.2	c: 2 x 2.00 = 4.0	
D: 2 x 0.50 = 1.0	d: 2 x 3.90 = 7.8	
15.0	13.6	1.0
Add 15% 2.3	2.0	0.2
17.3	15.6	1.2

We will assume that the quoted prices for timber delivered to the site are as follows:

38 mm x 100 mm — $1.30 per lin.m.
50 mm x 100 mm — $1.70 per lin.m.
38 mm x 200 mm — $3.00 per lin.m.

Thus the cost of materials per truss will be as follows:

38 mm x 100 mm timber: 17.3 @ $1.30 = $22.49
50 mm x 100 mm timber: 15.6 @ $1.70 = $26.52
38 mm x 200 mm timber: 1.2 @ $3.00 = $ 3.60
Sub-total $52.61
Add for nails   say   2.39
Cost of materials per truss $55.00

Labour

We will allow 10 hours for a carpenter with an unskilled assistant to manufacture each truss, and an additional 2 hours to position it on the roof. Thus the labour estimate will be based on 12 hours (1½ days) for one skilled and one unskilled man:

Labour cost per truss = 1½ x $ (15.00 + 7.50)
                                    = $33.75
Add materials cost as above   55.00
Thus unit rate for Item 5.03   = $88.75

5.04 No.16 gauge hoop iron straps (6 No.)

This is a minor item, and it is not worth wasting time on a precise estimate. An allowance of $2 for materials and $1 for labour should cover it, making a unit rate of $3.00.

5.05 50 mm x 100 mm purlins (88 lin.m)

Materials

The timber merchant has quoted a rate of $1.70 for 50 mm x 100 mm timber, so the materials rate will be:

Quoted rate per lin.m.   = $1.70
Add 15% allowance for cutting and
wastage   0.26
Sub-total $1.96
Add for nails   say   0.10
Total $2.06

Labour

We will allow 4 hours (½ day) for 2 carpenters plus 2 labourers to complete this task, i.e.

½ x $ [(2 x 15.00) + (2 x 7.50)] = $22.50

Thus the unit labour rate will be:

$(\frac{\$22.50}{88})$	= $0.26
Add materials rate as above	$2.06
Thus unit rate for Item 5.05	= $2.32

5.06 31 mm x 150 mm fascia board (36 lin.m)

Materials

Quoted rate per lin.m.		= $1.60
Add 15% allowance for cutting and		
wastage		0.24
Sub-total		$1.84
Add for nails	say	0.10
Total		$1.94

Labour

We will allow 5 hours for a carpenter (with a labourer to assist him) to complete this task, so the unit labour rate will be:

$\frac{1}{36} \times \frac{5}{8} \times \$\,(15.00 + 7.50)$	= $0.39
Add materials rate as above	$1.94
Thus unit rate for Item 5.06	= $2.33

5.07 Double doors 2.00 x 2.25 m (1 No.)

We will assume that the contractor has decided to sub-contract the fabrication of the doors and frame to a friend who has a specialist joinery workshop, and he has quoted an 'ex works' price of $95.00 to include the hinges and lock as specified. We will assume that the doors can be brought to the site as part of another load, and will charge $5 as a contribution to the overall cost of that load. Thus the materials cost 'on site' will be exactly $100.

Site labour will be required to fix the frame and doors in position, so we will allow 3 hours for a carpenter assisted by a general labourer, i.e.

$\frac{3}{8} \times \$\,(15.00 + 7.50)$	= $ 8.44
Add materials cost 'on site' as above	100.00
Thus unit rate for Item 5.07	= $108.44

This completes Bill No.5, and we can summarise the rates for the bill as follows:

Bill No.5 — Estimate Summary

Item	*Unit*	*Estimate ($)*
5.01	No.	42.81

5.02	lin.m	4.59
5.03	No.	88.75
5.04	No.	3.00
5.05	lin.m	2.32
5.06	lin.m	2.33
5.07	No.	108.44

Bill No.6 Roofing

6.01 Roof sheets as specified (85.8 sq.m)

In some countries components such as corrugated iron sheets are still made in imperial unit sizes (e.g. 3 ft x 5 ft, 3 ft x 10 ft) but are sold by the square metre. This makes estimating a little more complicated, but it still should not cause too much difficulty providing the correct conversion factors are used. The equivalent metric sizes for these sheets are illustrated below:

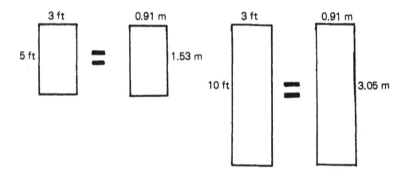

The roof surface that will have to be covered is sketched below:

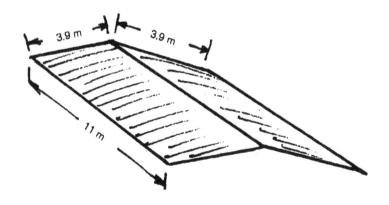

167

A single 10 ft long sheet will not be sufficient to cover the sloping face (3.9 m), so we will need to use one 10 ft plus one 5 ft sheet.

Next we need to work out how many sheets will be required to cover the full roof length of 11.00 metres. Since each pair of 5 ft plus 10 ft sheets must be lapped 1½ corrugations over the next pair, the effective width to be used for this calculation must be reduced from 0.91 m to 0.81 m. Thus the number will be:

$$\left(\frac{11.00}{0.81}\right) = 13.6; \text{ i.e. 14 sheet widths required.}$$

Thus we will need to buy a total of 28 No.10 ft long sheets and 28 No.5 ft long sheets.

There are four purlins on each side of the roof and each sheet width will be fixed to each purlin at the centre and at each side, so 12 screws and washers will be needed for each sheet width. Thus the total number of screws and washers required will be:

28 x 12	= 336
Add 10% allowance for wastage	34
Total	370

The merchant quotes the following prices:

Sheets:	$15.00 per square metre
Screws with washers as specified:	$ 0.40 each.

Thus the total cost of materials will be:

10 ft sheets: 28 x 3.05 x 0.91 x $15.00	= $1165.71
5 ft sheets: 28 x 1.53 x 0.91 x $15.00	= $ 584.77
Screws and washers: 370 @ $0.40	= $ 148.00
Sub-total	$1898.48
Add transport to the site: one round trip at the rate calculated under Bill No.1	40.00
Total 'on site' cost of materials	$1938 say

Labour

Allow 2 days' work for a team consisting of two skilled and three unskilled men:

2 x $ [(2 x 15.00) + (3 x 7.50)]	= $ 105
Add materials cost as above	$1938
Total	$2043
Thus unit rate for Item 6.01 = $\left(\frac{\$2043}{85.8}\right)$	= $23.81

6.02 Ridges 38 cm girth including fixing (11 lin.m)

We will assume that the ridging is supplied in 2 metre lengths at a cost of $3.00 per linear metre. (The cost of transport to the site is covered in Item 6.01 above.) Six pieces will be required, so the materials cost will be:

6 × 2 × $3.00   = $36.00

Labour cost: 2 hours for 1 skilled and 1 unskilled man:

¼ × $ (15.00 + 7.50)   = $ 5.63
Total $41.63

Thus estimated cost per lin.m Item 6.02
$= \left(\frac{\$41.63}{11}\right) =$ $ 3.78

The above calculation completes Bill No.6:

Bill No.6 — Estimate Summary

Item	Unit	Estimate ($)
6.01	sq.m	23.81
6.02	lin.m	3.78

Bill No.7 Painting and Glazing

7.01 Glazing to windows (6 No.)

A sub-contractor offers to glaze the windows at a cost of $45 each but, since this appears expensive, we will check the cost of carrying out the work with the contractor's own labour.

The merchant offers to supply glass at $15 per square metre. Thus costs would be:

Materials:		
Glass:	1.25 × 1.00 × $15 = $18.75	
Putty:	say	0.50
Clips:	say	0.50
Sub-total		19.75
Allow 20% cutting and wastage		3.95
Sub-total		23.70
Add cost of transport (part load)		2.00
Total		25.70

Labour: say 3 hours ($\frac{3}{8}$ day) for skilled man
@ $15.00   5.63
Unit cost for Item 7.01 using own labour $31.33

This is clearly more favourable than the sub-contractor's offer, and will be used as the basis for the rate for this item.

7.02 Paint external surface of walls (77 sq.m)
7.03 Paint external surface of walls (73 sq.m)
7.04 Paint windows (6 No.)
7.05 Paint doors and frame (1 No.)
7.06 Paint fascia boards (LS)

This particular contractor does not have any painters on his staff and, rather than recruit casual painters who may prove unreliable, makes a practice of sub-contracting all his painting work to a sub-contractor who he uses regularly and knows to be competent.

He has checked the sub-contractor's rates regularly in the past and knows that he is generally competitive, so in this case he passes a copy of this section of Bill No.7 to the sub-contractor. An allowance for overheads and profit will be made in the next chapter, but an allowance of 10 per cent for 'attendance on the sub-contractor' will be made at this stage to cover direct supervision and assistance, use of tools, etc.

Item	Sub-contractor's rate ($)	10% attendance allowance	Estimate ($)
7.02	5.50	0.55	6.05
7.03	4.50	0.45	4.95
7.04	6.00	0.60	6.60
7.05	15.00	1.50	16.50
7.06	20.00	3.00	22.00

Thus the summary of rates for Bill No.7 is:

Bill No.7 — Estimate Summary

Item	Unit	Estimate ($)
7.01	No.	31.33
7.02	sq.m	6.05
7.03	sq.m	4.95
7.04	No.	6.60
7.05	No.	16.50
7.06	LS	22.00

This completes the estimated unit rates for all seven bills, and the next stage will be to add an allowance for overheads and profit to each unit rate to give the full unit rates that will be entered in the tender. The considerations governing the decision on the percentage allowances for overheads and profit are discussed in the following chapter.

Chapter Seven

Bidding Policy

How Much Profit?

The estimate is a forecast — as factual and realistic as possible — of the actual cost of completing a particular job. The bid or tender is an offer to carry out the work for the client at a specified price and within a specified contract period. The difference between these two monetary figures — the estimate and the bid — represents the hoped-for profit margin for the contractor on that contract. Whenever he converts an estimate to a tender by making a decision on the level of profit to aim at, the contractor is faced with a major question — how much profit should he aim to secure?

No Easy Answer

Some contractors would say that the question is a ridiculously easy one to answer. Profit is a 'good thing' for a businessman and, since 'you can never have too much of a good thing' — you can never have too much profit! This is perhaps an understandable attitude, particularly in a high risk industry like building, but it does not get us very far. If a contractor builds unacceptably high profit margins into his bids, they will simply be undercut by his competitors and he will rapidly run out of work. So there is no easy solution to the question 'How Much Profit'.

Longer Term Approach

Every contract has its own peculiarities and poses its own problems. It is in effect a separate transaction with a single client. For this reason, most contractors are 'contract-oriented' rather than 'organisation-oriented' — which is perhaps a polite way of saying that they can't see the wood for the trees! It is this attitude which leads contractors to make the mistake of struggling along with a wildly fluctuating workload made up of occasional contracts (often carrying quite high profit margins), rather than setting — and keeping to — objectives that will maximise the overall profit of the firm as a whole over the course of a year.

Matching Resources to Workload

There is an ideal labour force, plant and equipment mix, stores inventory and managerial set-up for any particular type and size of workload. All contractors aim, efficiently or inefficiently, at achieving this. The problem is that most of the time they are trying to hit a moving target! Just as everything starts to work well — either an existing contract comes to an end and labour and resources are idle or a big new contract is awarded and the contractor has to 'steal' resources from existing jobs to keep the new client happy. Many contractors spend the biggest part of their time in a series of fruitless attempts to match their resources to their changing workload. They are not really managers, cooly using their judgement to achieve long term aims. They are jugglers, living from minute to minute and hoping not to drop anything too important!

Hire and Fire

The only way to match labour resources to a rapidly changing workload is to operate on a 'hire and fire' basis, ruthlessly taking on people in the knowledge that you will get rid of them as soon as work falls off. Casual employment is still a common feature in the building trade. It is certainly bad for the workmen involved, who have no chance of bringing home a regular pay packet for their families. But it is also against the real interests of the contractor too, because he ends up with a disloyal and unco-operative workforce. There will be no incentive for the site staff to push on with the contract, because they are intelligent enough to know they are working themselves out of a job. It is much more likely that they will contrive to take their time over the job, so as to put off the evil day when they are paid off. This means poor efficiency, high unit costs and poor profits for the contractor.

Using Sub-Contractors

Another common way of coping with a rapidly fluctuating workload is to make extensive use of sub-contractors. A 'labour only' sub-contractor takes on parts of the contract (other than the supply of materials and sometimes plant and tools) at a fixed price, and it becomes his problem to complete the work productively and cope with the labour force. But this solution is not as easy as it sounds. Not all sub-contractors are reliable, and if they let you down — you get the blame from the client. There is also the possibility that they will waste —

or even steal — materials. Finally, if you have too many sub-contractors on the job, controlling and co-ordinating them all can be a nightmare for the site supervisor. They tend to get in each other's way, and claim 'extras' for difficulties caused by other sub-contractors. A further point is that it is quite common for contracts to contain a condition forbidding sub-contracting 'the whole or any part of the work without the express permission of the client's representative'.

A Steady Workload

There are two ways of matching resources to workload. The first, outlined above as the way of the juggler, is to accept that the workload will always go up and down as a fact of life and do one's best to change the resources of the firm to suit what is on hand. The other approach, the way of the manager, is to understand that it is very difficult to run an efficient business if manpower and other resources are changing all the time. This implies approaching the problem from the other direction, and seeing if there is a way of ironing out the fluctuations in the workload.

The Way of the Manager

Every building firm, with its managerial and technical skills and physical resources, has its own unique potential profit ceiling — but few get anywhere near achieving it. The main reason for this is that it is so difficult to arrange a reasonably steady workload in the building business, so that everyone and everything is fully employed all the time. To get into this happy position, the firm needs enough offers of work from potential clients to be able to pick and choose jobs that will occupy the workforce and the available plant and financial resources to the best advantage. This in turn requires a conscious emphasis on marketing, discussed more fully in an earlier chapter, plus a readiness to manipulate tendering policy to suit the express objective of maintaining a steady (or perhaps steadily expanding) workload.

Tendering Policy

When a firm starts out in business, it has to chase every enquiry and bidding possibility really hard and keep profit margins low to make it worthwhile for the clients to give them a chance. The client will be accepting a risk in awarding work to an untried firm, and will probably only do so if the price is very competitive and there is a promise of really good service. As the firm expands and becomes better known, it can obtain better profit margins and even (politely!) refuse to tender for work that looks too risky or will overload the order book. It will have definite targets for orders and will reduce profit margins when work is short, and ease them up as things improve.

Work Load

Thus the first general aim in setting a tendering policy is to try to maintain a steady workload. Every firm has an optimum workload, at which all its resources are fully utilised, but none are overstrained. 'Too little work' is a common complaint among contractors. But overtrading can be just as dangerous, particularly if financial resources are limited.

Market conditions

But tendering policy must also take account of current market conditions. At times when work is scarce, it may be necessary to go for work further than the normal working limit or of a different type to that usually constructed. When estimating for such a job, the contractor must take special care that his cost forecasts properly reflect the unusual working conditions.

Marketing

Since building is a 'boom or bust' industry, in which the market conditions fluctuate so violently, a further general factor in tendering policy is the need for a contractor to build up his reputation. Marketing techniques were discussed in the first chapter, but one way for a contractor to become well known is deliberately to seek out prestige projects that will bring him into the public eye.

174

Market Conditions

Prestige

Thus the contractor may be influenced by an opportunity to demonstrate his skill and ability to potential clients, by taking on a prestige building job even at a lower than usual profit margin. However, before taking on such a project, he must be absolutely confident of his ability to execute it efficiently as news of failure travels faster than news of success.

Three aspects

Thus there are three aspects to general tendering policy which will be kept in mind by the contractor when deciding on the level of profit margin to seek:

1. *Workload:*

Aiming at a steady workload at a level that would make optimum use of available resources.

2. *Market Conditions:*

The general level of profit margins in the industry will be affected by the intensity of competition for available work.

3. *Prestige:* .

Much less important than the other two considerations, but there can be occasions where a prestige contract yields a tangible marketing advantage.

Building at a profit

Most contractors seek to build at a profit — and the unit rates put forward in the completed bill of quantities will put an upper limit on the profit that can be earned on a particular

job. Of course even that profit will have to be worked for on the site, but no miracles of operational efficiency can make up for bid rates that are below actual unit costs. Few contractors have escaped the occasional experience of taking on work at a loss, and regretting their over-optimistic approach at the tendering stage. Whether he is running his own private business or has been put in charge of it by members of his family who provided the initial capital or is responsible to outside shareholders, the primary duty of the manager remains the same. It is to ensure a reasonable return on the money invested in the business — by building at a profit.

What is profit?

Profit for the contractor is made up of three separate components:

(a) PERCENTAGE RETURN ON MONEY INVESTED, RELATED TO INTEREST RATE ON BORROWED MONEY.

(b) PREMIUM TO ALLOW FOR CONTRACTUAL RISKS — DIFFICULTY OF JOB, FINANCIAL STANDING OF CLIENT, WEATHER RISKS, ETC.

(c) REAL PROFIT, DEPENDS UPON:
    1 POLICY.
    2 EFFICIENCY.

Even after we have allowed for direct costs in our basic estimated unit rates and added a percentage to cover site overheads and general overheads, the margin we are left with is not all *real* profit.

Percentage return

If the owners of the firm had not got their money tied up in a building business, it is not likely that they would put their cash in a big sack and hide it under the bed. They would put it to work for them in some way. Even in the deposit account at the bank it would earn a positive return. If it was lent to a trustworthy local businessman who needed capital to expand his retail shop into a supermarket, it would earn a much higher return. Thus, before the business can be said to be earning a real profit for its proprietors, it needs to earn enough to 'pay for its keep' in terms of the sort of percentage return that the shareholders could enjoy if their capital was invested elsewhere.

What return?

The percentage return that would be looked for depends on a number of factors. If the capital is all provided by the proprietor and building is the only business he knows, he might be satisfied with the 5 or 6 per cent that the local bank pays on money invested on deposit. But a new business may be started with the help of borrowed money, perhaps partially financed by a bank overdraft or loan at 12 or 13 per cent. In that case a return at that level must be looked for before the business starts to earn a real profit. It is a hard fact that money costs money, and the businessman who borrows at 15 per cent to earn 5 per cent will not remain a businessman for very long!

Return on what?

There is a difference between *return on capital* and *return on turnover.* This is important in the contracting business where stage payments and credit from suppliers mean that a 6-month $25,000 contract can be financed with fixed and working capital of perhaps only $10,000. If a second $25,000 contract follows on in the next six months, this means that capital of $10,000 has financed $50,000 of annual turnover. Thus, even if interest on capital costs 15 per cent per annum, it can be fully recovered by a return of one-fifth of that (3 per cent) on work done.

Risk premium

Building will always be a high risk business, and money invested in a building business stands a much greater chance of being lost than money in a bank account. The next chapter discusses safety precautions, insurance and other ways of 'taking care', but even these can only reduce the degree of risk rather than eliminate it. Builders must live with risk and, with the best will — and the highest level of efficiency — in the world, every now and again things will go wrong.

What are the risks?

Some risks — such as a spell of unseasonal bad weather — cannot possibly be foreseen at the tendering stage. Others, such as the specification of materials known to be in short supply coupled with a short contract period and a tough penalty clause for late completion, can and should be foreseen and allowed for in setting bid prices. In this way the contractor is acting as his own private insurance company and charging the client a 'special insurance premium' for the additional risk that he is being asked to run. The main four risk areas that need to be taken into account in deciding on the element of profit that will relate to a risk premium are:

1. The contract
2. The site location
3. The client
4. The architect (or engineer)

The contract

Are there any unusual provisions in the contract documents? Are any unusual materials or construction techniques called for? Is the contract period tight — and are there penalties for

late completion? Are the drawings clear and unambiguous? What are the provisions for interim and final payments — and retention money? What are the provisions for arbitration in the case of a dispute?

The site location

Is the site within a reasonable distance of the office? A site visit will certainly be necessary before a realistic estimate can be prepared. A contractor who knows that bad ground conditions can be expected should decide whether his experience (or that of his staff) is sufficient to allow the contract to be carried out at a reasonable profit.

The client

Have you worked for this client before? Does he make payments promptly on the issue of certificates by the architect? Will he stick to the original design or change his mind continuously resulting in stoppages of work and extra costs that are difficult to recover? If you have not encountered him before in business dealings, make discreet enquiries through friends or your bank manager.

The architect

If you have not carried out work designed by this architect, it is often sensible to make enquiries as to his reputation for dealing fairly with claims and other difficulties which may be encountered.

Anticipated real profit

What is left over after a percentage return on capital invested and a notional insurance premium to cope with the probable level of risk is the anticipated real profit. Real profit provides a source of funds to strengthen and expand the business, and the level of real profit a contractor is able to earn is also an important indicator of operational competence and efficiency. A contractor who is able to show a record of profitable operations will be judged a good risk by potential lenders, since it is likely that a good return will be earned on any funds that they provide and interest and principal will be repaid promptly.

A policy decision

The level of profit to seek is a policy decision, and will be decided on when the estimate has been prepared. Note that it is *not* the job of the estimator to prepare the tender or bid. The manager must make the final decision on the profit level to include, for in a well-run firm policy decisions are decided by those who have the most knowledge of the firm's operations and bear most responsibility — namely the management.

The bid equation

The eventual tender rates are decided by the bid equation:

$$\text{ESTIMATE} + \text{OVERHEADS} + \text{PROFIT} = \text{TENDER}$$

Both estimated unit rates and overheads are based on facts — they can only be altered if the firm's operating methods and structures are altered. Thus the profit decision by the manager is the key to the actual tender price, and that decision will be made in the light of the strength of the order book and of the resources available to the business.

The master chart

The contractor should keep a master chart showing contracts in progress and accepted tenders. This must be updated regularly — at least once a month. By consulting this the manager will be able to judge whether his company would be able to cope with the additional work within its current resources of finance, manpower and plant and equipment.

Financial resources

Inadequate financial resources would be the most serious consideration, since men can be recruited and additional plant and tools can be bought if funds are available. If there is any doubt about whether sufficient capital is available to finance the additional work, a cash flow projection should be prepared to identify and quantify the possible shortfall. At this stage the bank manager should be consulted as to whether he would be prepared to advance temporary funds to allow the contract to be undertaken. This may appear a waste of time when the bid might well be unsuccessful, but it might save the embarrassment of having to withdraw an accepted tender due to inadequate funds or — much worse — a financial crisis and possible bankruptcy in the middle of an otherwise successful contract.

The example

We will now return to the example that we have been dealing with in the last two chapters. We now have a bill of quantities and a set of *estimated* unit rates which will require an addition to cover overheads and profit. It is quite possible to make different percentage additions on different items, and it sometimes makes sense to put the biggest part of one's profit margin on the items that will be completed first to assist with cash flow. The danger with such an uneven distribution is that the contract might be varied by the client or his architect, reducing the quantities on the high profit margin items and increasing those with low margins. Thus we will take the safe route and decide on a fixed percentage addition and apply it to every item in every bill.

Forecast of total estimated cost

Before we decide what percentage addition to apply, we need an approximate forecast of our total estimated cost as a basis from which to work. This could be done by going through the estimated unit rates on all the bills and getting a total of the estimated direct costs for each item in each bill. It is also necessary to make a forecast of the percentage addition that we might eventually come up with. Thus if the total of estimated costs came to $12,000 and the percentage addition was 20 per cent, the tender price would be $14,400. The forecast does not need to be absolutely accurate, and an experienced contractor might take a short cut and make a direct forecast of the total estimated cost (and make an appropriate correction later if he was too far out). We will take the short cut and forecast the total estimated cost as $12,000 and the tender price as $14,400. To simplify the example, we will base recovery on the gross estimated cost, including preliminaries, although in practice these are normally deducted and the same total recovery is achieved by a slightly higher percentage charge against the remaining bill rates.

Overheads and profit

Our percentage addition will cover overheads and profit and will be made up from five components as follows:
1. Site overheads
2. General overheads
3. Return on capital
4. Allowance for risk
5. Real profit.

1. Site overheads

We have already covered most of our site overheads in the preliminaries bill, including all the items illustrated below (except for the site signboard, which is a minor item that will be found from the allowance for plant, tools and vehicles).

The only site overhead that we have yet to cover is the cost of site supervision. On larger contracts there is often an item in the preliminaries bill against which the contractor is required to provide for the cost of a full-time site agent or site foreman to control the work. But in this case we will

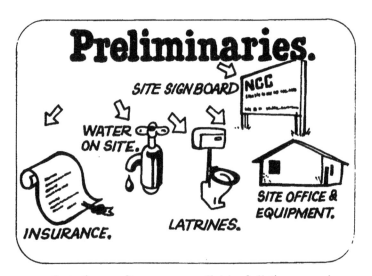

Preliminaries.

SITE SIGNBOARD

WATER ON SITE.

INSURANCE.

LATRINES.

SITE OFFICE & EQUIPMENT.

assume that the craftsmen are reliable full-time employees, and the cost of overheads can be trimmed by providing only for part-time supervision by a foreman who will also be responsible for other contracts.

We will allow for 3 weeks of full-time supervision, 3 weeks of half-time supervision and 6 weeks quarter-time supervision by a foreman who is paid a wage one-third higher than a typical craftsman, so his 'all-in' cost will be $20 per day or $100 per week.

Thus cost of site supervision for the duration of the contract will be:

$$[3 \times (½ \times 3) + (¼ \times 6)] \times \$100 \qquad = \$600$$

This site overhead of $600 must be recovered from all the bill items on a proportionate basis. The total estimated cost is $12,000, so this means that site overheads will represent:

$$(\frac{\$600}{\$12000}) \times 100 \qquad = 5 \text{ per cent}$$

2. General overheads

General overheads will cover all those costs that cannot be directly allocated to individual contracts. They will include a salary for the contractor, the costs of running his car or pick-up, the costs of running an office or yard, audit, accountancy and legal costs and all the other miscellaneous payments and fees that have to be met in order to run a successful business. The only way in which these costs can be recovered is from the revenue that accrues from the interim and final payments on contracts carried out by the business. For accounting purposes we look on every contract as a little mini-business on its own, with its own incomes and expenditures. The overall organisation of the contracting firm is effectively there to provide a service to all these mini-businesses — and naturally that service has to be paid for. The fairest way of charging out the general overheads, which represent the cost of this service, is by making a percentage charge to each contract on the basis of either contract value or total estimated cost. Thus a $20,000 contract will be expected to 'pay' twice as much as a $10,000 contract, and a $40,000 contract will be expected to pay twice as much as a $20,000 contract.

The best way to deal with forecasts of percentage additions needed to recover general overheads is to operate a general system of financial planning and budgetary control for the enterprise (for further details see the companion volume *Financial Planning for the Small Building Contractor*).

We will assume that the firm's administration budget (covering all general overhead items) for the year is $18,000. The other figure we need is the annual production budget, which gives the forecast of the work that will be completed (i.e. the total of *direct* costs on all contracts) by the firm during the course of the same financial year. For the sake of the present example, we will take this figure as $300,000.

It is our intention to earn an extra $18,000 on this $300,000 worth of work to cover general overheads, which implies an oncost of $18 on every $300 of work. It will be most convenient to have this in the form of a percentage, i.e.

$$(\frac{\$18,000}{\$300,000}) \quad = \quad \textbf{\textit{6 per cent}}$$

This means that the contract we are tendering for will (if we get it!) contribute $720 (6 per cent of $12,000) to general overheads. This is one twenty-fifth (4 per cent) of the total overheads of $18,000, which is correct since the contract will be one twenty-fifth of the total production budget.

3. Return on capital

We need to recover the cost of capital tied up in financing the contract. This will depend on the terms of payment laid down in the contract documents, plus an assessment of whether the client will be as good as his word in honouring these commitments and making payments promptly. To obtain a really accurate idea of the cost of capital, it is necessary to make a cash flow projection for the contract (the cumulative difference between money expended on the work and income received from the client).

On contracts where mobilisation advances are available, the cost of capital to the contractor will be quite small (since he is using his client's money rather than his own). At the other extreme, if a contract made no provision for any form of interim payment, the contractor would have to cover the complete costs of financing the project from his own or borrowed funds. Most contract terms lie between these two extremes, so the experienced contractor will be able to make a fairly accurate assessment of the cost of financing a project for which he is tendering, and leave the preparation of a detailed cash flow projection until the contract has been awarded.

In the present example, we will assume that the cost of capital will work out to one-third of the contract sum and that this will have to be available for an average period of 4 months. The cost of capital for this firm is 15 per cent per annum.

Our forecast of the contract sum was $14,400, so the cost of capital will be one-third of this or $4,800. It will be needed for four months (i.e. one-third of a full-year), so the equivalent annual cost would be $1,600. The cost of finance is 15 per cent per year, so the actual cost will be:

$$\$1,600 \times (\tfrac{15}{100}) \qquad\qquad\qquad = \$240$$

This must be recovered from a total estimated direct cost of $12,000, so the percentage addition to cover return on capital will be:

$$(\tfrac{\$240}{\$12,000}) \qquad\qquad\qquad = 2\ per\ cent$$

4. Risk

Risk must normally be assessed rather than calculated, and some of the risks to which the contract will be subject have already been covered as insurance premiums and various allowances in the estimates for bill items. In this case we will assume that the contractor decides that a risk factor of *3 per cent* would be a realistic provision against those risks that have not been covered elsewhere.

5. Real profit

The allowance for real profit will also have to be assessed

rather than calculated. Ideally we would like as large a slice of real profit as possible while remaining the lowest bidder! But we have no way of knowing what the rival bids will be, so it will be a question of making a policy decision on the level of profit margin to seek bearing in mind how anxious we are to obtain the contract. Generally it is best to have a steady policy on the size of profit margin to seek, and vary this slightly upwards or downwards according to the state of the order book. In this case we will go for a real profit margin of *4 per cent.*

This may seem a very low profit, but remember that we have already allowed for the cost of interest on capital and all general overheads (including a salary for the contractor himself). What is more, the job will be over in 3 months — so it should be possible to earn similar profits on other work during the remaining 9 months of the year.

The reader may not be convinced, and a common objection is 'Why not double the profit margin to 8 per cent. Even if the workload halves from $300,000 to $150,000 the contractor still gets a profit of $12,000 — and only half as much worry!' This may seem an attractive argument on the surface — but what about the recovery of general overheads? On $150,000 of work, using our recovery rate of 6 per cent, we will only recover $9,000 worth of general overheads. The remaining $9,000 will have to come out of profit, so the 'apparent' real profit of $12,000 will come down to $3,000 (unless the contractor takes a cut in salary!).

Total percentage addition

We have now decided on allowances for overheads and profit as follows:

1. Site overheads   — 5 per cent
2. General overheads   — 6 per cent
3. Return on capital   — 2 per cent
4. Allowance for risk   — 3 per cent
5. Real profit   — 4 per cent
   Total percentage addition = *20 per cent*

The effect of this 20 per cent addition to estimated unit rates is set out in the following table, and the bid rates in the final column will be the rates that we will insert in the bills of quantities and use to calculate the tender sum.

Calculation of Bid Unit Rates

Item	Estimate $	Addition (20%) $	Bid rate $
1.01	235	47	282
1.02	164	33	197
1.03	120	24	144
1.04	45	9	54
1.05	91	18	109
1.06	180	36	216
1.07	80	16	96
1.08	INCL.	—	INCL.
1.09	INCL.	—	INCL.
1.10	181	36	217
1.11	150	30	180
1.12	100	20	120
1.13	45	9	54
2.01	0.47	0.09	0.56
2.02	3.75	0.75	4.50
2.03	5.00	1.00	6.00
2.04	0.94	0.19	1.13
2.05	2.81	0.56	3.37
2.06	4.99	1.00	5.99
2.07	2.53	0.51	3.04
2.08	3.90	0.78	4.68
2.09	18.14	3.63	21.77
2.10	2.14	0.43	2.57
2.11	30.00	6.00	36.00
3.01	7.44	1.49	8.93
3.02	9.38	1.88	11.26
3.03	94.15	18.83	113.00 say
3.04	94.15	18.83	113.00 say
3.05	94.15	18.83	113.00 say
3.06	0.88	0.18	1.06
3.07	1.10	0.22	1.32
4.01	32.76	6.55	39.31
4.02	16.00	3.20	19.20
4.03	27.00	5.40	32.40
4.04	2.86	0.57	3.43
4.05	3.09	0.62	3.71
4.06	3.93	0.79	4.72
4.07	3.25	0.65	3.90
5.01	42.81	8.56	51.37
5.02	4.59	0.92	5.51
5.03	88.75	17.75	106.50

Item	Estimate $	Addition (20%) $	Bid rate $
5.04	3.00	0.60	3.60
5.05	2.32	0.46	2.78
5.06	2.33	0.47	2.80
5.07	108.44	21.69	130.00 say
6.01	23.81	4.76	28.57
6.02	3.78	0.76	4.54
7.01	31.33	6.27	37.60
7.02	6.05	1.21	7.26
7.03	4.95	0.99	5.94
7.04	6.60	1.32	7.92
7.05	16.50	3.30	19.80
7.06	22.00	4.40	26.40

Completing the tender

We have now at last reached the stage where the tender can be completed. This will be done by inserting the bid rates in the blank bill of quantities printed at the end of Chapter Five. Where the unit is LS (Lump Sum), the bid rate can go straight into the 'amount' column. Otherwise the bid rate should be entered in the 'rate' column. The tender amount for each item is calculated by multiplication:

QUANTITY x RATE = AMOUNT

e.g. *Item 2.01*   162   x   0.56   = $90.72

Totals for each bill

Each bill should be added up, and the sub-total transferred to the final 'SUMMARY SHEET'. The last-but-one stage in the tendering process is to add up the separate bill sub-totals on the summary sheet to obtain the tender price.

Do-it-yourself?

The reader may wish to work through the calculations himself using the blank bills and summary sheet in Chapter Five. He can then check his calculations against the completed bill of quantities printed on the following pages.

Item No.	Description	Unit	Quantity	Rate	Amount
	BILL No.1 PRELIMINARIES				
1.01	Allow for all necessary plant, tools and vehicles for carrying out the work	LS			282.00
1.02	Provide temporary site office for the duration of the contract complete with table, chairs and all necessary office equipment and requisites.	LS			197.00
1.03	Provide temporary shed for secure storage of goods, materials and components	LS			144.00
1.04	Allow for preparation of access roads, tracks and hardstandings for the delivery, transport and storage of bulky materials	LS			54.00
1.05	Provide temporary fence for the duration of the contract as specified	LS			109.00
1.06	Allow for providing a supply of water for the works	LS			216.00
1.07	Allow for provision of all facilities, including latrines, to comply with statutory safety, health and welfare regulations in respect of all work-people employed (including employees of subcontractors)	LS			96.00
1.08	Allow for transport of workpeople to the site and any other disbursements arising from their employment	LS		INCL.	—
1.09	Allow for expenses of watchmen and other security measures as may be deemed necessary to protect the works	LS		INCL.	—
1.10	Provide temporary scaffolding for the proper execution and completion of the works	LS			217.00
1.11	Allow for obtaining 'Contractor's All Risks', public liability, employer's liability and other insurance cover in accordance with the terms and conditions of the contract	LS			180.00
1.12	Allow for provision of contract performance bond to the value of 10 per cent of the contract sum in accordance with the terms and conditions of the contract	LS			120.00
1.13	Allow for removing all rubbish and debris from the site and cleaning the building both internally and externally prior to handing over to the client	LS			54.00
	Transferred to summary sheet				1,669.00

Item No.	Description	Unit	Quantity	Rate	Amount
	BILL No.2 GROUNDWORKS				
2.01	Site clearance and removal of topsoil to approx. depth 10 cm	sq.m.	162	0.56	90.72
2.02	Excavation to lower level of hardcore over building area, av. depth 10 cm	cu.m.	7	4.50	31.50
2.03	Excavation in foundation trenches, av. depth 90 cm	cu.m.	22.5	6.00	135.00
2.04	Prepare bottom of trench for concreting	sq.m.	12.5	1.13	14.13
2.05	Return, fill and ram excavated material around foundation	cu.m.	17	3.37	57.29
2.06	Place and compact hardcore 20 cm deep within building and blind with gravel to receive concrete	sq.m.	53.8	5.99	322.26
2.07	Adjust levels, supply and roll gravel path 10 cm min. thick around building	sq.m.	52.5	3.04	159.60
2.08	Excavation for access road, av. depth 30 cm	cu.m.	13.5	4.68	63.18
2.09	Supply and compact hardcore to access road, min. 30 cm deep	cu.m.	13.5	21.77	293.90
2.10	Supply and roll as specified gravel surface to access road 10 cm thick	sq.m.	45	2.57	115.65
2.11	Level surplus soil as directed and tidy site on completion	LS			36.00
	Transferred to summary sheet				1,319.23
	BILL No.3 CONCRETE				
3.01	Formwork to sides of foundation base	sq.m.	12.5	8.93	111.63
3.02	Formwork to sides of foundation walls	sq.m.	62.4	11.26	702.62
3.03	Concrete 1:2:4 to foundation base	cu.m.	2.5	113.00	282.50
3.04	Concrete 1:2:4 to foundation walls	cu.m.	6.2	113.00	700.60
3.05	Concrete 1:2:4 to floor slab	cu.m.	4.1	113.00	463.30
3.06	16 mm diameter mild steel rods in top of foundation walls including laps, bends and tying wire	kg	102	1.06	108.12
3.07	8 mm diameter mild steel rods at 30 cm centres in both directions in floor slab including all tying wire, distance blocks and spacers	kg	142	1.32	187.44
	Transferred to summary sheet				2,556.21
	BILL No.4 BLOCKWORK AND PLASTERING				
4.01	Concrete block walls 20 cm thick in cement mortar	sq.m.	72.3	39.31	2,842.11
4.02	Lintels over windows 20 x 25 x				

Item No.	Description	Unit	Quantity	Rate	Amount
	140 cm reinforced with 4 No.6 mm diameter mild steel rods	No.	6	19.20	115.20
4.03	Lintel over doorway 20 x 25 x 240 cm reinforced with 4 No.6 mm diameter mild steel rods	No.	1	32.40	32.40
4.04	External wall plastering 17 mm thick, to receive paint	sq.m.	77.0	3.43	264.11
4.05	Internal wall plastering 17 mm thick, to receive paint	sq.m.	73.0	3.71	270.83
4.06	Floor screed, average thickness 3 cm	sq.m.	53.8	4.72	253.94
4.07	Supply, position and grout in place ½" anchor bolts for roof fixing	No.	30	3.90	117.00
	Transferred to summary sheet				3,895.59

BILL No.5 CARPENTRY AND WINDOWS

Item No.	Description	Unit	Quantity	Rate	Amount
5.01	Supply and fix 1.00 x 1.25 steel windows as specified	No.	6	51.37	308.22
5.02	Timber 100 mm x 100 mm wall plates, including fixing to anchor bolts	lin.m.	35	5.51	192.85
5.03	Supply and fix roof trusses in accordance with details supplied	No.	3	106.50	319.50
5.04	No.16 gauge hoop iron straps 32 mm wide and 40 cm long including screw fixing to wall plates	No.	6	3.60	21.60
5.05	Supply and fix 50 mm x 100 mm purlins	lin.m.	88	2.78	244.64
5.06	Supply and fix 31 mm x 150 mm fascia board	lin.m.	36	2.80	100.80
5.07	Double doors 2.00 x 2.25 m including frame, hinges, lock and fitting, all in accordance with details supplied	No.	1	130.00	130.00
	Transferred to summary sheet				1,317.61

BILL No.6 ROOFING

Item No.	Description	Unit	Quantity	Rate	Amount
6.01	Corrugated galvanised mild steel roof sheets fixed to timber with ¼" diameter galvanised roofing screws 2½" long each with one diamond-shaped bitumen washer and one galvanised steel washer. Side laps of 1½" corrugations	sq.m.	85.8	28.57	2,451.31
6.02	Ridges 38 cm girth, including fixing as specified	lin.m.	11	4.54	49.94
	Transferred to summary sheet				2,501.25

BILL No.7 PAINTING AND GLAZING

Item No.	Description	Unit	Quantity	Rate	Amount
7.01	Glazing to windows as specified	No.	6	37.60	225.60

Item No.	Description	Unit	Quantity	Rate	Amount
7.02	Paint external surface of walls as specified	sq.m.	77	7.26	559.02
7.03	Paint internal surface of walls as specified	sq.m.	73	5.94	433.62
7.04	Paint windows as specified	No.	6	7.92	47.52
7.05	Paint doors and frame as specified	No.	1	19.80	19.80
7.06	Paint fascia boards as specified	LS			26.40
	Transferred to summary sheet				1,311.96

SUMMARY SHEET

Bill No.	$
1. Preliminaries	1,669.00
2. Groundworks	1,319.23
3. Concrete	2,556.21
4. Blockwork and Plastering	3,895.59
5. Carpentry and Windows	1,317.61
6. Roofing	2,501.25
7. Painting and Glazing	1,311.96
	14,570.85

Did you get it right?

Did you get it right? I hope so. But even with simple arithmetic, we can all make mistakes. And arithmetical mistakes in preparing a tender can cost a contractor a lot of money. This is why the completion of the summary sheet was described as the last-but-one task.

Final stage — Check your calculations!

The final stage of the tendering process is to check carefully through all the earlier calculations to make certain that there are no mistakes. If possible, someone other than the person who did the original calculations should do the checking. In a small firm this may not be possible — but if you can't find anyone else, please don't take this as an excuse for not checking your own calculations. Every time you sign your name at the bottom of a tender document, you are committing yourself to a solemn legal undertaking that may either make or lose you a great deal of money. The best advice for a carpenter is 'measure twice and cut once'. The best advice for a contractor is 'check twice and tender once'.

Check in reverse order

Here is a small tip that may be useful if you have to check

your own calculations. *Carry out the check computation in reverse order.* If you first multiplied the quantity by the unit price, check by multiplying the unit price by the quantity. If you first added the unit amounts from top to bottom, check by adding up the page from the bottom to the top. By altering the order of the calculation, you can often pick up a mistake that might otherwise pass unnoticed.

Contingencies

One final point is that no allowance has been made for contingencies (things that might happen but which we have no way of knowing about at the time that the contract is prepared). Note the difference between contingencies and provisional items such as 'extra over for excavation in rock', where we know what might happen but not its extent. Contingencies are allowed for by a single percentage addition (usually 10 per cent) to the total of all bill items. For our example this addition would be $1,457.09, making the total bid *$16,027.94.* The contractor must remember that no part of the contingencies sum will be released to him without an express written justification of the extra work or variation concerned by the client's representative.

Chapter Eight
Taking Care

A High Risk Business

The building contractor does not need to be told that he has chosen to compete in a high risk business. The fact is that building and civil engineering contracting companies head the list of bankruptcies and liquidations in almost every country. A building contractor faces many risks, such as:

— rising prices on fixed price contracts
— delays due to bad weather
— carelessness by his men
— clients defaulting on their debts
— building up a pool of labour and plant only to find that the building programme is cut or the contract cancelled.

Risks Cannot be Eliminated

Whatever happens the contractor will never eliminate all the risks. His client pays him to take risks, as well as to use his skills. The client wants a building and he wants to know how much it will cost *before* construction starts.

By signing a contract, the client ensures that the cost to him will be known, and the risk will be passed on to his contractor in return for the margin of profit allowed.

But Risks can be reduced

Although he cannot eliminate all risks, there are ways in which he can reduce them. The main way is by his own foresight, by planning and forecasting, so that snags can be dealt with before they grow into disasters. But he can also reduce risk by an intelligent use of·insurance — a subject that is discussed later in this chapter.

What is 'an accident'?

When things go wrong, there is a tendency to say 'it was an accident', as if that is sufficient explanation — because 'accidents will happen'. It is true that every contractor, no matter how experienced he may be, sometimes make mistakes. But some contractors are much more accident-prone than others. For them the law of averages appears to have been suspended, and one way or another something always seems to be going wrong.

Plato's definition

Over 2,000 years ago, Plato defined an accident as 'that which happens blindly without intelligent design'. If we accept this definition, perhaps we should be less complacent in saying 'accidents will happen'. After all the contractor is paid by his client to use his intelligence, skills and experience to implement the project effectively — so that what happens on his site should happen as a result of 'intelligent design'. Thus before we go on to discuss insurance cover to defray the cost of accidents and risks of various kinds, we should first examine ways in which the contractor can apply sound management principles to reduce the likelihood of such misfortunes occurring on his jobs.

Risk Management

Since the building contractor is in the risk business, he cannot ignore the concept of risk management in which the possibility of losses of various kinds is consciously considered and evaluated in advance with a view to minimising the possible damage to the organization as a whole. This takes time and effort, but pays off by ensuring that staff become risk-conscious and carry out their tasks in a way that cuts down the chances of loss and also, when the occasional lapse does occur, ensures that there are remedial actions and procedures already thought out which can be applied to minimise its effect.

Four Approaches

Once a risk has been identified and evaluated, there are four ways of dealing with it which can be described as risk avoidance, risk retention, risk transfer and risk reduction:

Risk Avoidance

Some risks can be simply avoided altogether as a matter of policy. A fixed price tunnelling or trenching contract in an area where the ground conditions are known to be difficult is bound to involve very considerable risks, even if the contractor is able to secure apparently generous unit rates. Unless the contractor has very strong financial reserves and complete confidence in his ability to deal with the technical and operational problems that will arise, such a contract is better avoided. Risks of default by private clients can be avoided by confining one's workload to public bodies or requiring prepayments or deposits of funds with trustworthy third parties. The risk of robbery can be minimised by making transactions by cheque whenever possible, and encouraging senior staff to open bank accounts so that they can be paid by cheque. In the field of physical accident prevention, the best way of stopping people falling down deep trench and foundation excavations is to complete the construction work and backfill to ground level as quickly as possible, so that such an accident simply cannot occur.

Risk Retention

Some risks cannot be avoided, and building contracting is by its very nature a risk business — and this has to be accepted provided the risks are known and quantified in advance. Some can be covered by suitable insurance policies, but there are other risks that no insurance company will cover. The most obvious example is that it is not possible to pay a small premium to secure a policy that will guarantee a profit on the next contract that is awarded! (Although it is possible to cover 'loss of profits' in certain cases, such as a joinery workshop being burned down with the result that outstanding orders cannot be completed.) Some risks have to be passed on to an insurance company by law and others are normally required according to the terms of the contract, but there is sometimes some room for discretion on the type of cover. For example, the law normally requires motor vehicles to be covered for 'third party' risks, but the policy holder may decide that it would be prudent to take out a 'comprehensive' policy. Large firms with strong financial backing can afford to shoulder more of their own risks, but smaller firms have a greater need for full insurance cover since a single loss could wipe out their limited working capital and send the business into liquidation or bankruptcy. The real danger is risk retention by default, in which a contractor — like a man sleeping peacefully in the path of a cyclone — has no clear idea of the risks he is running, until it is too late to do anything about them!

Risk Transfer

The most common form of risk transfer is by taking out an insurance policy, which has the effect of ensuring that the costs or losses resulting from the occurance of certain defined risks will be met by the insurance company rather than by the policy holder. The conditions of the policy and the nature of any exclusion clauses should be read very carefully to ensure that the contractor knows precisely which risks he has transferred and which he has retained. A 'cheap' policy may turn out to be expensive in the long run if the insurance company fails to meet the claim due to an overlooked exclusion clause in the fine print on the back of the policy. There are other ways in which a contractor can transfer his risks under the contract. One way is by sub-contracting a part of the work, although it is not possible for a main contractor to evade his own direct responsibility to the client for errors and omissions on the part of his sub-contractors. Another example of risk transfer is a policy decision to hire plant rather than buy it outright, so that breakdowns and repairs are paid for by the plant hire company in return for a fixed hourly, daily or weekly charge.

Risk Reduction

Whether risks are retained or transferred, it makes sense for any businessman to do his best to reduce them to a minimum by the pursuit of sound policy and operational procedures. Because building is a high risk business, it is also a business in which accidents are common. Sometimes these accidents result in injury or even loss of life. Other accidents merely hold up work on the site or result in damage to property. All these accidents fit in with Plato's definition of 'that which happens blindly without intelligent design' and all are costly in terms of both time and money. And for every actual accident, there are usually several 'near misses' which could have been serious but were avoided at the last minute by a stroke of good fortune. So the scope for risk reduction in most building businesses is usually very considerable. There is no natural law that builders should be accident-prone and, despite the riskiness of the industry as a whole, some contractors regularly achieve their planned and budgeted performance and have excellent safety records. Most accidents do not just happen. They are caused. And the root cause is usually some form of human ignorance or neglect.

Safety pays!

Apart from the questions of legal and contractual liability, a positive attitude to accident prevention and site safety backed by a modest investment in training, protective clothing and equipment will pay off well for the contractor in terms of higher productivity and fewer hold-ups. Injury accidents do not just result in misery and pain for your employees and their families, although this should be an important factor

196

for the caring employer. The knowledge that one of their friends and colleagues has been badly hurt is bound to adversely affect the morale of your labour force, and unhappy people are unlikely to give that 'little extra' productive push that brings a contractor a reputation for high speed, quality project implementation. There is an established trend in many countries for more construction workers to organise themselves into trade unions and union leaders, in addition to their concern for rates of pay and conditions of employment, rightly emphasise safety and welfare. Thus a firm with a poor safety record may find itself with a strike on its hands, and be forced into providing safety measures without gaining the goodwill that would have been achieved if it had done so voluntarily. There will also be demands for generous compensation to the injured man and his family.

Health and Safety Legislation

Ministries of Labour in numerous countries are promoting legislation affecting health and safety at work and, once it is enacted, backing it up with regular visits to factories and building sites by qualified factory inspectors. These inspectors usually have considerable powers, including the ultimate power to shut down a site completely if the dangers are unacceptably great or if earlier warnings have not been heeded.

An Accident Prevention Programme

One way of emphasising the commitment to safety is to set up an accident prevention programme, aimed at reducing the number of accidents that occur. All accidents should be recorded so that they can be analysed, and lessons learned for the future. If the employee (or his family) should later make a claim against the company, the courts are likely to take a serious view if the site supervisor failed to make a proper record in an accident book. Even apparently minor incidents may give rise to later complications, and all foremen and site agents should be given strict instructions to record all accidents without exception. As the records build up, the contractor will begin to get an idea of which types of accident occur most commonly and will bias his accident prevention programme to tackle these risks as a first priority.

Common Accidents

Some of the most common accidents occur through a mixture of ignorance and plain carelessness. Workers do not look out for protruding nails, splinters, ragged edges or oil spillage, and this emphasises the need for 'good housekeeping' on the construction site. A tidy site is usually an efficient site, and a few hours of a general labourer's time spent removing nails from used timber, stacking surplus materials and clearing away and disposing of rubbish results in quicker and safer access around the site for the remainder of the work force, besides giving a more workmanlike impression to the casual visitor.

Lifting and Carrying

Manual workers spend a lot of their time lifting and carrying materials and equipment around the site, and many of these items are heavy and awkward to handle. Serious and lasting back injuries can be caused by incorrect lifting procedures, and the simple rule should be 'bend your knees rather than your back' so as to avoid putting a bending stress on the spine. Workers should not be required to lift loads that are heavier than they can manage comfortably, since it is usually more economic as well as safer to use a team of two or three men and finish the job more quickly. When lifting is being done by teamwork in this way, it is necessary to have one person in charge to give a lead as to how and when the item is to be lifted, moved and placed so that the load is always evenly shared.

Protective Equipment

Some foolish workers feel that it is in some way 'unmanly' to wear safety helmets or other protective clothing. The contractor should not indulge or accept this attitude, since it exposes him to potential loss as an employer just as it exposes the employee to danger. Thus site supervisors should be instructed to ensure that appropriate safety equipment is made full use of, and workers who refuse to conform should be subject to dismissal.

Ladders

Faulty ladders and the incorrect use of ladders are common causes of accidents on building sites. Ladders should be regularly inspected and faulty rungs should be replaced. Ladders should *never* be painted, since the skin of paint might hide a dangerous defect which will cause a labourer to fall when carrying a heavy load. Ladders should be placed on firm ground at an angle of 1:4 to the wall as illustrated below, and it is a wise precaution to anchor the toe and secure the top of the ladder so that there is no danger of slippage. Short ladders should *never* be balanced on oil

drums, upturned buckets or piles of bricks to give a little extra reach.

Temporary Works and Scaffolding

Although the architect or engineer is responsible for the design of the main structure, contractors are usually responsible for the design of temporary works. Even if their calculations are examined by the consultant, they will remain responsible for any deficiencies in the event of failure. Temporary structures to support reinforced concrete formwork need special care, since the loads resulting from the placing and compaction of the wet concrete can be very considerable and, should the formwork start to move, the eventual collapse could have very serious consequences. The fixing of scaffolding is a job for specialists rather than unskilled labourers, and the completed scaffolding should be thoroughly inspected before it is used. As with ladders, timber used in scaffolding should be left unpainted so that any defects can be clearly seen, Particular attention should be given to joints and wall fixings, supports and anchorages. Guard rails should be provided along all walkways, and toe-boards should also be fixed to stop items being accidently kicked off high scaffolds.

Hoists, Lifts and Suspended Scaffolds

Ropes, chains and lifting appliances that are used for support or vertical movement should be regularly inspected by a competent and qualified person, and are often subject to special regulations. Such items normally require special

Scaffolding Dangers.

insurance cover, but this in no way absolves the contractor from responsibility for seeing that all such items are kept in good working order. Since the workman who uses such equipment cannot easily tell whether or not it is fully operational, the law takes a very serious view of contractors who permit their equipment to deteriorate into a dangerous condition and, in the event of an accident, the contractor may face criminal charges as well as heavy compensation to the casualties and the direct losses resulting from the accident.

Plant and Machinery

Plant and machinery should be operated strictly in accordance with the manufacturer's instructions, and regularly maintained by a competent and trained mechanic. The tendency for some employees to remove safety screens and guards to speed up or ease operations must be strictly checked, since this much increases the risk of serious accidents and may even invalidate the insurance cover. The temptation for employees to ride as passengers in insecure positions on dumpers, etc. often leads to injuries, and site supervisors should be instructed not to allow it under any circumstances.

Excavations

Many building accidents occur during the excavation of trenches and foundations. Earth is the most variable material with which the contractor has to deal, and untimbered trenches may stand for weeks in one part of the site but collapse without warning quite close by. This book is not intended to introduce the reader to the theories of soil mechanics, some of which are quite complicated and depend upon accurate measurements of the soil in question. However, every contractor should have a reasonable grasp of the nature, properties and dangers associated with the use of the main materials with which he has to work and the following few paragraphs are intended to set down a few of the main 'dos and don'ts' as a brief guide to the contractor. But the golden rule in carrying out earthworks is to make sure that any errors are on the safe side, since collapses occur with very little warning and can be catastrophic for workers who are trapped.

Excavation near existing buildings

Additional care must be taken when excavating in the vicinity of existing buildings, since the new construction may endanger their stability. If in doubt, the architect or engineer should be approached for instructions as to the need for sheet piling or other protective measures.

Inspection

In many countries there is a legal requirement that all trenches, excavations and tunnels should be inspected by a competent person at least daily, and a written and signed record of the inspections and examinations that have been carried out should be kept on the site. If an accident should occur, the record will be taken as evidence and the person making the inspection may be cross-examined to determine his competence and experience.

Watch the edge!

The edges of excavations are the parts most likely to crumble or collapse, so the first sensible rule is not to stand too close to them! The second thing to remember is that any superimposed load on the edge of an excavation results in a stress equivalent to deepening the trench, so excavated material and pipes, etc. should be stockpiled or stacked well back from the edge. Where a mechanical digger is used, it should also be kept away from the edges of open excavations, particularly where men are working underneath. If there is doubt about the stability of the excavation and sufficient space is available, it may pay to step back the sides of the excavation to reduce the weight of overhanging soil.

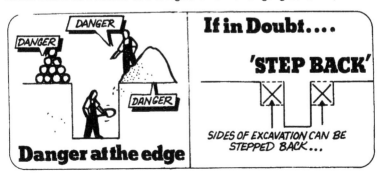

Danger at the edge

If in Doubt....

'STEP BACK'

SIDES OF EXCAVATION CAN BE STEPPED BACK...

Access to Trenches

Access to trenches should be by ladders, which should overhang the edge of the trench by at least 1 metre so that workers, supervisors and inspectors can step well clear of the edge when getting onto or off the ladder. Proper timber gangways should be provided for people to cross trenches at reasonable intervals, and areas between access points should be securely fenced. In public roads or rights of way, special precautions must be taken to ensure that dangerous excavations are clearly fenced and signposted and the full length of excavation should be illuminated during the hours of darkness.

Drainage

For maximum stability, trenches should be well drained by maintaining a gentle slope and digging a small sump at the lowest point. Where there is a danger of the trench flooding, more ladders and access points should be provided so that it could be evacuated rapidly if an emergency arose.

Timbering

Deep trenches (and shallow trenches in treacherous soil) should be properly strutted and braced by competent and experienced people. The materials should be carefully inspected before installation, and the completed timbering should also be inspected to ensure that all struts and braces are fully secured. It is also important that alterations and

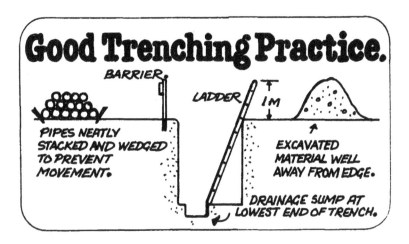

Good Trenching Practice.

BARRIER.

LADDER

1M

PIPES NEATLY STACKED AND WEDGED TO PREVENT MOVEMENT.

EXCAVATED MATERIAL WELL AWAY FROM EDGE.

DRAINAGE SUMP AT LOWEST END OF TRENCH.

dismantling should be done by a competent crew. The following illustration gives an indication of appropriate timbering techniques, but it must be emphasised that this is a job for an expert.

First Aid

Even if the highest standards of safety procedure are followed, accidents will still occur occasionally so every site should at least have a first aid box readily available to provide prompt treatment. On larger sites a member of the staff should be designated as first aid officer and should receive appropriate training in accident procedures. First aid boxes should be kept properly supplied and should be clearly marked 'FIRST AID' and should not be used for any other purpose than the storage of first aid materials and dressings. The telephone number and address of the nearest doctor and the nearest hospital should be kept readily available so that no time will be lost in obtaining professional medical attention.

Welfare

One aspect of 'taking care' which also makes a contribution to site morale and therefore to potential levels of productivity is the provision of basic welfare facilities for site staff. Basic requirements are an adequate supply of water for both drinking and washing, sanitary facilities and a shelter for storing clothing and personal belongings and taking meals. Some contractors begrudge spending money on even these basic facilities, but are repaid with a slovenly and uncaring workforce. The excellent rule of 'do as you would be done by' applies here, and the contractor would be well advised to

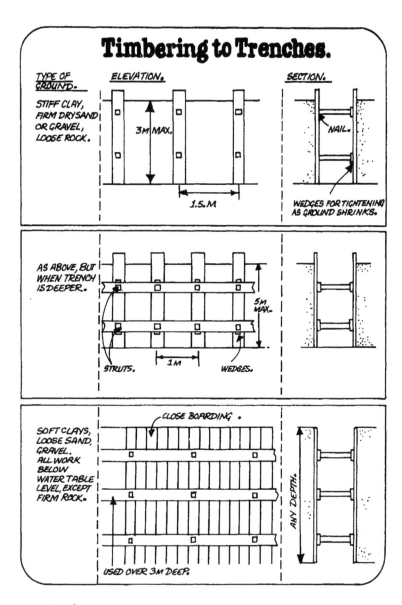

Timbering to Trenches.

TYPE OF GROUND.	ELEVATION.	SECTION.

STIFF CLAY, FIRM DRY SAND OR GRAVEL, LOOSE ROCK.

3M MAX.

1.5.M

NAIL.

WEDGES FOR TIGHTENING AS GROUND SHRINKS.

AS ABOVE, BUT WHEN TRENCH IS DEEPER.

5M MAX.

STRUTS.

1M

WEDGES.

SOFT CLAYS, LOOSE SAND, GRAVEL. ALL WORK BELOW WATER TABLE LEVEL, EXCEPT FIRM ROCK.

CLOSE BOARDING.

ANY DEPTH.

USED OVER 3M DEEP.

spend a few minutes pondering how he would like to be a manual worker on one of his own sites. If the answer is 'not much' or 'that's why I left my old boss to start on my own', he should consider improving things before he in turn starts to lose his best workers.

Safety is everyone's business!

The contractor and his site supervisors should give a lead in safety matters and, as usual, the most important element in giving a lead is setting a good example themselves by wearing protective equipment and not indulging in dangerous practices. But safety is also the business of everyone on the site, and everyone on the site should be encouraged to be 'safety-minded'. Communication of the safety message can take many forms, but simple illustrated posters — strategically sited — can provide useful reminders provided they are relevant and not overdone. Some useful standard notices are illustrated below:

Damage Control

Not all accidents result in injury to people. The contractor's balance sheet may also suffer from accidents which damage or destroy his plant, materials or equipment. These non-injury accidents can take a large bite out of contract profits and can only be reduced by ensuring that the work force is both knowledgeable and caring. Operators of mechanical plant and equipment must be carefully chosen and well-supervised, since a few minutes' carelessness, such as failing to wash out the concrete mixer at the end of the day, can cost a completely disproportionate sum to put right. The wise contractor will ensure that his accident reporting and costing system also embraces these non-injury accidents, so that procedures and instructions can be progressively clarified and tightened and the same mistakes are not repeated time after time.

Fire

A fire in an almost completed building can cost big money. Even if there is adequate fire insurance cover, the client will not get his building on time and the contractor may get a reputation as being 'accident-prone' and lose business as a result. It is wise to consult the experts from both the insurance company and the local fire service on preventative measures, fire fighting equipment, fire safety and emergency measures for employees. There is the added advantage that this advice is normally provided free!

Theft and Fraud

Another aspect of 'taking care' that unfortunately cannot be ignored is that of theft and fraud, both by the contractor's own employees and by third parties. The normal reaction by most contractors who think they know their own business is that these things are common enough in the industry as a whole 'but I know my sites and I know my people, and it could never happen to me'. This may be true. But it is more likely that it is not. A graphic example of what can happen every day and yet be revealed only by chance is given by Earl Lorence, director of security for the Metropolitan Tobacco Co., New York, in the book *Internal Theft and Investigation Control*:

> 'Some years ago, the head of a giant plant engaged in government contract work wanted a publicity picture of thousands of employees coming out of the main gates at the end of a shift. To set the scene, the gates were locked and camera men were stationed at the watchmen's hut. The rumour was that the FBI was going to search them all for stolen goods and parts.
> 'When the picture had been taken, the gates were opened and the workmen left for home. On the floor where they had been standing were over 4,000 items of tools, parts, soap, towels and even a 15 pound sledge hammer.'

Naturally the less you own, the less there is to steal! But the loss of $200 worth of tools and equipment during the course of a $10,000 contract is as serious for a small contractor as would be the loss of $20,000 to a much larger firm on a million dollar project. In each case, if the profit margin on the work was 10 per cent, the elimination of this theft would push the firm's profits up by as much as 20 per cent.

The Risk

The builder has a bigger security problem than most businessmen, because his sites are more accessible than most factories and workshops and, in many countries, there is a ready black market among unscrupulous competitors for stolen materials and tools. Even quite large items of mechanical plant, such as concrete mixers, are sometimes stolen and repainted so that they are very difficult to trace. Other items, such as timber, blocks, bags of cement and roof sheets appeal to the dishonest 'do-it-yourself' enthusiast, and there have been examples of complete houses being constructed from bricks stolen from a single large building site!

What to do about it?

The first stage in tackling any problem is to find out all you can about it. So a good documentation and cost-recording system is a must. You need to know the number of bags of cement that were allowed for in the estimates for each site on a stage by stage basis, so that these can be used as base figures with which to compare actual usage. If actual usage is more than the estimate, there are two possible reasons. One is theft and the other is wastage. Both come straight out of your profits and therefore straight out of your pocket, so either way you will need to find out the reason why the figures do not tally. Your accounting and cost recording system should be effective enough to highlight which items are disappearing from which sites, and this will help you to know what sort of protective measures to take with a view to safeguarding items that are most attractive to the thief.

Stock Control

Stock control in the stores needs careful attention, with all issues simply but clearly recorded and care being taken to ensure that people employed in the stores are completely trustworthy and are paid a wage or salary that is in keeping with their responsibility for your goods and your money. A regular stocktaking and reconciliation with issues, deliveries and returns will help to ensure that any discrepancies are quickly traced.

Physical security measures

The nature of physical security measures on sites will depend on the location as well as on the value of the materials and equipment and their attractiveness to a potential thief. Location can be a crucial factor. In a village where everyone knows everyone else, things may be quite safe stacked on an unfenced site. But in the large anonymous city, gangs of criminals are always on the look-out for easy pickings and much tougher measures are called for. In urban areas, sites should normally be securely fenced or boarded and valuable items should be placed in sturdy sheds under lock and key.

Vandalism is also a problem in some cities, and it may be necessary to consider the additional overhead of employing watchmen (possibly with guard dogs) for protection outside working hours. For detailed advice on what security measures would be appropriate for a particular site, the local police force should be consulted. Police officers are normally happy to give this advice, because crime prevention measures limit the scope for the criminal and make their job a little easier.

Perk or pilferage?

Pilferage and theft by one's own employees is a delicate problem. Some employers accept that their employees can have odd scrap materials from their sites that are of no value to the firm. Craftsmen are sometimes allowed to borrow tools for the week-end to work at home or on private jobs. The clerk would not be branded as a criminal if he absent-mindedly walked home with a pencil in his pocket. The problem is one of custom and degree. If borrowing of tools and taking of surplus materials is an accepted 'perk' of being employed with your contracting firm, this is quite acceptable providing employees know just *how far they can go.* The clerk may take one pencil. But what about five? Or fifty? The line has to be drawn somewhere.

Be consistent

If such perks are to be allowed at all, the contractor should make clear rules on just how far they go, and also state quite definitely that to go beyond them will be regarded as a criminal offence. Application of the rules should be absolutely consistent between one employee and the next, and senior staff should be careful to set a good example. If clear standards are set and applied consistently and uniformly, internal dishonesty can be controlled. If not, the problem will degenerate to the point where people will not only steal a few bags of cement but will also 'borrow' one of your wheelbarrows to make the task easier!

Insurance

By taking some of the precautions outlined above, risks can be significantly reduced, but the building contractor runs a risky business in a risky world, so they cannot be completely eliminated. Thus risk transfer must come into the picture, and the usual method of transferring risk is to take out an insurance policy with an established insurance company who will guarantee to reimburse the costs involved if the risk does materialise in return for a fee known as the 'premium'. Like the contractor, the insurance company is in business to make money, so the higher the risk the higher will be the premium charged for covering it. The accident-prone contractor who makes claim after claim will find that quotations for future premiums will rise steeply, while his more careful competitor with a better claims record will be able to cover his risks more cheaply.

Definition
One legal definition of insurance is 'a contract in writing whereby one party, called the INSURER agrees in consideration for either a single or a periodical payment, called the PREMIUM, to indemnify another party, called the INSURED, against loss or damage resulting to the INSURED on the happening of certain specified event or events.'

The 'Fine Print'
The above definition gives a flavour of the sort of complex legal phraseology that abounds in a typical insurance policy, although it is only fair to say that the better companies are doing their best to simplify the language within the limits required for clarity in the event of a dispute. Even so, many insurance policies remain hard to read and hard to understand. Thus many busy contractors fail to find the time to read their policies carefully enough, and confidently believe themselves to be covered for all eventualities when some are definitely left out due to various 'exclusion clauses' contained in the fine print on the back of the policy. If a costly accident comes under the heading of one of these excluded risks, the company will not pay and the resulting loss could bankrupt the firm.

Know the questions
Insurance is a specialised field and it would not be realistic to expect the contractor to know all the answers. But he does need to understand the risks that he is running, and the sorts of insurance policy that are available from various companies to cover them. Then, even if he doesn't know all the answers, he should at least know the right questions to ask so that he gets cover that fits his needs and is most cost-effective in relation to the funds that he has available.

Types of insurance
The remainder of this chapter discusses the range of insurances available to the contractor, their purpose and some of the points that need attention when seeking quotations

and deciding on the most appropriate forms of cover. The discussion covers insurance under the following seven broad classes:

1. Vehicle Insurance
2. Employers' Liability Insurance
3. Public Liability Policies
4. Plant Insurance
5. Contractor's All Risks' Policy
6. Other Policies
7. Contract Performance Bond.

1. Vehicle Insurance

This is probably the most widely accepted and understood form of insurance cover, and vehicles cannot be legally operated in most countries without some form of basic cover against damage or injury to third parties. Apart from breaking the law, the contractor without satisfactory cover may well end up facing huge third party or passenger liability claims involving sums that far exceed his capacity to pay. It is wise to extend beyond the statutory minimum cover to a 'comprehensive' policy that will cover damage to the vehicle itself. Although the premiums for comprehensive cover are inevitably larger, they are not very great in comparison to the overall annual cost of keeping a vehicle on the road. Furthermore, they give the contractor more confidence in budgeting and estimating for vehicle costs, without having to keep a large sum of money in reserve to cope with emergencies.

Keep them on the - Road.

WITH COMPREHENSIVE COVER..

2. Employers' Liability Insurance

If any employee of a contractor suffers accident or injury as a result of carrying out his work for the firm, then the firm can be held liable to pay compensation. As the Trade Unions grow in power and importance they will, quite rightly, try to ensure that any such compensation to one of their members is as substantial as the law allows.

In some countries Employers' Liability (or Workmen's Compensation) Insurance is required by law. Even if this is not the case, it may well be a condition of contract, as the client will want to be protected against the possibility of any claim ultimately devolving on him should a claim against the contractor not be met in full. The standard condition of contract also requires that the workmen's compensation insurance be extended to cover all employees of sub-contractors, both nominated and direct, who may be engaged from time to time on the site.

The caring contractor will wish to know that in the event of a major catastrophe involving a large number of workmen, there will be an immediate source of finance from an insurance company to meet legitimate claims for damages and loss of earning capacity. No monetary sum can really compensate for death or permanent injury, particularly of the father of a young family, but cash would certainly be needed and a satisfactory insurance policy will ensure that it will be provided. Besides humanitarian considerations, it is good commercial sense to ensure that such an accident cannot cause a dent in the firm's cash flow and funds will be available to get on with the work and recover the progress that has been lost.

Of course, having an insurance policy does not mean that the contractor can forget about safety measures on his sites. In fact the policy will probably require the contractor to take

211

all reasonable precautions, and a representative from the insurance firm may visit the site to ensure that this is being done. Insurance should be seen as an additional safeguard for the good employer, not as a let-out for the careless or incompetent manager.

3. Public Liability Policies

The building contractor is a rather unusual kind of business-man. Most of his work is carried out on other people's property and, in the case of the civil engineering firm, may be carried out in the open air on a site to which the public has full access. Thus his work can very easily give rise to claims from third parties who may stray onto the site, or suffer damage or injury in some indirect way.

For example, careless excavation can give rise to all sorts of claims. Water pipes or electricity cables can be broken, giving rise to claims for repair from the service concerned. Excavation close to an existing building can cause subsidence, and the owner would have good grounds for a damages claim. If a trench excavation is not properly protected, a passer-by could fall in and be injured. If a tall building is being constructed in a town centre, a block may fall accidentally and damage a car parked below.

SOME BUILDING WORK INEVITABLY BRINGS THE PUBLIC INTO RISK

In some cases the damages can be very considerable. Suppose an eight storey office block is nearing completion in a busy city centre. If there is a foundation failure in the middle of a large concrete pour, the collapse of scaffolding, formwork and wet concrete onto the street below may result in a traffic accident and several fatalities among both drivers and pedestrians. Obviously all these innocent people (or their families) have a claim against someone, and the unfortunate contractor (even if he is not really at fault) is always first in the firing line.

212

But in a strict legal sense, the claim could equally (or better) be made against the Client or Principal as the ultimate owner of the structure and the contractor is merely his nominated agent for the performance of the work. It is partly as a result of this possible ultimate liability of the Principal himself that most contracts stipulate that the contractor should maintain adequate insurance as protection against claims which may be made by members of the public for injury or loss or damage to property which may arise during the course of the contract. Since claims may be addressed to either the contractor or the client (or, to be on the safe side, to both!), it is normally a requirement that the liability policy is taken out in the joint names of the Principal and the Contractor. Thus any claims that are received will be dealt with together under the same policy without any need to dispute how the liability should be shared between the parties.

Since the liability policy is partly taken out to directly benefit the Principal, it is quite common for the cost of the premium to be dealt with as a specific item in the 'Preliminaries' bill.

4. Plant Insurance

On projects where a substantial amount of plant and equipment is employed, this will have to be covered by a special policy. This may well be made a condition of contract on certain civil engineering jobs, where the Principal may make substantial advances on an account basis to enable the contractor to equip himself to carry out the work. These 'mobilisation' advances will be safeguarded by an additional clause to the effect that plant or equipment on arrival on site is deemed to be in the joint ownership of the Principal and the Contractor. Naturally the Principal will wish not only to protect his own capital payment (which will have been made in advance) but also to make sure that funds would be available to permit the repair or replacement of any plant which may be written off or damaged during the course of the construction programme.

Mobilisation advances are rarer on building contracts, since they do not normally require so large an investment in plant and equipment. Thus the joint ownership clause is not required. However, there is still likely to be a condition that the contractor maintains full insurance cover on his machinery and equipment, both in respect to replacement or repairs and for liabilities to third parties.

Methods of cover vary according to requirements and the nature of the risk. The plant and equipment itself may be covered directly in the Contractor's All Risks policy that is discussed below. Liability to third parties will be dealt with either under the general liability policy or by means of a special extension under a plant or equipment policy.

5. Contractor's 'All Risks' Policy

When a builder signs his name to a contract document, he becomes liable to hand over the building in good condition at the end of the job and to carry out any repair work up to the end of the maintenance period. All building contracts place some responsibility on the contractor for the safety of the works, and most require that he should specifically insure against serious — although hopefully unlikely — risks which should endanger them. These serious risks, which are usually described graphically by the insurance companies as *perils,* could lead to catastrophic damage or even total destruction of the building when it is almost complete. Thus the cost of renewal and repair could far exceed the accumulated retention money held on the client's behalf.

Few contractors are in the fortunate position of being backed by sufficient working capital to be able to afford to rebuild the structure from their own resources, and many would be simply forced into liquidation or bankruptcy if the client sued them in the courts in an attempt to enforce the contract. Insurance against such 'perils' therefore must also make sense from the client's point of view, since he could otherwise be left with a collapsed building and a failed contractor but no source of funds to put things right.

When a Contractor's All Risks policy is in operation, these unlikely but potentially crippling risks are taken over by the insurance company, which is able to spread its risk over a large number of contracts for which it writes insurance and build up a sufficient fund by quoting a quite low premium on each individual contract — usually less than 1 per cent of the value of the contract. Out of every thousand contracts which it covers, two or three may lead to claims and the accumulated fund will be available to make good the damage or, in the event of total loss, the original contract price would be available to start the job again from the beginning. Because both contractor and principal are at risk, the insurance policy is usually taken out in their joint names and is normally included in the bill for 'preliminaries' so that the premium can be directly priced by the contractor.

The 'perils' that must be guarded against on a standard building contract are Fire, Lightning, Explosion, Storm, Tempest, Flood, Bursting or Overflowing of Water Tanks, Apparatus or Pipes, Earthquake, Aircraft or other aerial devices or anything dropped therefrom, Riot and Civil Commotion. Civil engineering contracts often go even further and require cover against all forms of loss or damage except that caused by War, Occupation by the Principal, or the Engineer's design of the works. Thus an All Risks Policy really covers most risks rather than all risks, but it is still a very valuable way of protecting the interests of both the contractor and the client. Naturally the policy does not cover faulty workmanship by the contractor or the use of defective materials.

The risks that remain to be covered by the insurance company are wide ranging, and include fire, weather risks, flood, subsidence and collapse, toppling of plant, theft, vandalism and accidental damage from almost any cause. What is more, the cover runs from the moment when the first materials are delivered to the site, through the completion of the structure to the end of the maintenance period. Thus the insurer will want to examine the contract documents, and his quotation will be based on a declaration of the value of the works, plant and materials to be insured which will be provided by the contractor on a standard form (see Appendix 1 below). If the contract price varies during the course of the work, an adjustment is made at the end of the contract period on a proportional basis either by an additional payment from the contractor or by a partial return of premium by the insurance company.

If the insurance company is in any doubt about the right premium to quote, it may send an insurance surveyor to the site to assess the risk more accurately. He will be particularly interested in the soil conditions, the risk of flooding and the security precautions to be taken on the site. The insurance company may lay down specific security measures that will have to be installed by the contractor as conditions for cover against theft and vandalism — which may extend to the type of sheds to be used for storage of tools and materials and the make of locks to be used to secure them.

There is an onus on the contractor to declare any special features of the contract or any unusual plant or techniques that he proposes to employ, so that the insurers can make a completely fair assessment of the risks that they are offering to run on his behalf. A typical 'declaration of value' form is illustrated below:

Specimen: Declaration of Value to be Insured

Principal:
Contractor:
Contract:
Period:

1. The whole of the Contract Works, whether permanent or temporary, including materials incorporated or to be incorporated therein. $

2. Temporary buildings not included in (1) above whilst at the site (s) specified or in transit by road or inland waterway within the territorial limits specified. $

3. Contractor's plant, equipment, tools and tackle (not otherwise insured) all the property of the Assured or for which they may be responsible. $

4. Spare parts, fuel, stores, provisions and all other property of a similar nature held or located at the contract site (s) including personal effects of employees not exceeding _ for any one loss. $

5. On costs and expenses necessarily incurred in removing debris or the portion or portions of the property insured by Items 1 and 2 above, destroyed or damaged by any peril insured against.

6. On Architects', Surveyors', and Consulting Engineers Fees. $

7. Escalation/inflation reserve value:
   Item No.1 $) This declaration is
   Item No.2 $) subject to acceptance
   Item No.3 $) by Underwriters
   Item No.4 $) $

   Total Sum Insured $

When the quotation has been accepted, the contractor will receive a cover note of the kind illustrated opposite and this will have to be produced to the client's representative before work starts on the site.

The rate of premium charged will depend upon the nature of the work and the competence and claims experience with the contractor. Plant, tools, temporary buildings and temporary works are normally separately rated. Excesses are applied in order to avoid the overheads of dealing with claims for minor losses, and it is usually preferable for the contractor to accept these rather than pay the higher premiums that would have to be charged for 100 per cent protection.

Specimen: Cover Note

INSURED		A/C
COVER	Contractor's All Risks and Liability Insurance.	
PERIOD	from to	
PRINCIPAL		
CONTRACT		
SUM INSURED	Contract works, materials etc – $ Site improvements – $ Plant/Equipment) – $ Spare parts, fuel etc.) – $ Removal of Debris – $ Professional Fees – $ Total Sum Insured – $	
MAINTENANCE PERIOD		
LIABILITIES	Limit any one claim – $ Any one period of insurance – Unlimited	
DEDUCTIBLES	Storm, flood, subsidence etc. – $ All other Claims excluding fire/lightning – $ Third Party Claims – $	
RATE		
PREMIUM		
SECURITY		A/C

CLASS OF POLICY	DATE OF ISSUE	COVER NOTE NO:	PREMIUM POLICY CHARGE TOTAL	

Contractors who carry out a large number of building contracts of a standard type may find it worthwhile to negotiate their 'All Risks' cover on a yearly basis. The policy might cover 'Any contract for the erection, alteration or repair of buildings not exceeding three storeys in height and built of brick, stone or concrete and roofed with tiles, concrete, metal or asbestos'. The policy would then cover all jobs falling within this definition up to an agreed contract price, and the premium would be based on the estimated annual turnover with a proportionate adjustment to the actual figure for completed work at the end of the year.

Contracts falling outside the agreed definition must of course be covered separately.

6. Other policies

There are many other contingencies against which it may pay the contractor to insure although, as stated earlier, unfortunately no builder can insure against failing to make a profit on a contract! It is certainly sensible to insure your head office, office equipment and files against damage or loss, and small contractors operating from home may be able to arrange this cover as an extension to their house 'contents' policy. Most insurance companies will discuss and advise on any special requirements or problems faced by the individual businessman. The important thing is to consult them *before* the risk is run — no one will issue a retrospective cover note after the worst has come to pass!

REMEMBER ALSO TO INSURE YOUR OFFICE AND ITS EQUIPMENT

7. Contract Performance Bond

This is really a form of insurance although it is designed to cover risks to the client rather than to his contractor. The value is usually 10 per cent of the contract sum, and it is to be made available to the client in the event of the contractor failing to complete the project. The client usually requires the bond to be in the joint names of the contractor and some form of major financial institution, such as a bank or an insurance company.

The performance bond is effectively a guarantee of the contractor's competence. For a small building contract one would expect a 10 per cent performance bond, but for work of a more complex nature (such as a dam or a suspension bridge) or where the contract period is unduly lengthy, the performance bond figure may possibly be as high as 25 per cent of the contract price.

In the United States of America, there is a rather different approach to bonding. The figure often starts at 100 per cent of the contract price and reduces as work is carried out. All this of course, is in addition to the 10 per cent retention money which a contractor has to forfeit until the maintenance period is completed. The purpose of the performance bond is not so much a penalty should the contractor fail to perform the work, as a sum of compensation payable to the principal which will assist him to bring a new contractor on to the site should the original contractor fail for any reason to complete the work which he has contracted to do. The expenses incurred by a principal in the event of failure by a contractor would be the cost of preparing new documents, quantities and estimates and seeking tenders in preparation for a new contractor to take over and finish the job. Inevitably these bidders will quote higher prices than the original contractor, as it is always expensive to move onto an existing site, correct earlier mistakes and take over where someone else has left off (or given up!).

The negotiation of suitable performance bonding arrangements is often extremely difficult for the small new contracting firm, since the guarantor knows that a substantial sum of money will be at risk for months or even years and it is likely to be completely lost if the contractor fails to perform as promised in his bid. Thus the established contractor with a good record for completing work promptly starts with a very large inbuilt advantage.

Performance bonds were originally seen mainly as financial guarantees, and were therefore generally provided by financial institutions such as banks and finance companies. However, these institutions are really in the business of short term lending, rather than the assessment of long term risks and the giving of guarantees. Thus they have become more reluctant even to quote for the provision of bonding facilities which can give rise to long-lasting and substantial contingent liabilities, particularly on projects that may last several years and be followed by a one year maintenance period during which they remain at risk.

To some extent, the insurance market has been able to meet the demand for this type of guarantee. But only where a contractor can demonstrate a relatively sound financial position and has assets over which a charge can be taken in support of the guarantee.

By taking out a performance bond the contractor does *not* pass the risk over to the guarantor. He is still first in the line of potential losers, since the guarantor will only step in if he fails. Moreover he signs away some of his rights to resist a later claim which he may see as unfair. The Principal will be able to call up the guarantee immediately the contractor is seen to be in default of his contractual obligations, *in the eyes of the client's professional advisers and for any reason whatsoever.*

The method of charging for the provision of a performance bond is a fee related to the size and nature of the contract and assessment of the experience, competence and financial stability of the contractor. Besides payment of the fee, the contractor is likely to have to assign part or all of his assets as collateral in support of the bond. A contractor with a poor record, or even one with a record too short to be satisfactorily assessed, may not be able to obtain a bond even at a disproportionately high fee. This is due to the generally adverse claims experience on construction performance bonds, which are seldom generators of worthwhile profits for either banks or insurance companies but do involve them in considerable risks over which they have little or no control.

If a bond cannot be obtained, the only alternative for the contractor is to pay over a cash sum equivalent to 10 per cent of the contract figure as a deposit with the client in evidence of good faith. Since this deposit has to remain for the complete duration of the contract and maintenance period, this will make a very considerable dent in the firm's working capital. Besides the cash flow implications, profits will themselves be reduced by the amount of interest payable if the sum has to be borrowed or the interest forgone on what would otherwise have been a positive cash balance.

Responsibility

The objective with all the different forms of insurance discussed above is to transfer risks of various kinds from the shoulders of the contractor to those of his insurer. But it is not just a question of shuffling off responsibility and trouble to somebody else. The contractor always retains responsibility for the safe and steady progress of work on the site, and he will be held responsible by his client for anything that goes wrong. He cannot insure against his own incompetence (with certain very limited exceptions), and a number of policies specifically allow the insurance company to reclaim costs that result from inadequate or foolish actions by the contractor or his employees or sub-contractors. Even if a current policy is not made void by such actions, the contractor who wants to stay in business should remember that the insurance company does not have to cover his next contract or renew his annual policy at current premium rates or even at all. Thus it is wise to maintain a good business relationship with them.

A good record

Insurance companies trade for profit, just like the contractor himself. If a contractor has a record of irresponsibility and makes many frivolous claims on one or more insurance companies, the word will soon get round. The rule for the man who seeks insurance is similar to that for the man who seeks a loan — build up a reputation for competence and integrity by being fair and reliable in your dealings with others. This is something that can only be achieved over the years, but it can be lost overnight by a single foolish or unscrupulous action. Just as bankers are happy to loan money to the 'good credit risk', insurance companies will offer their most competitive premium rates to the 'good accident risk'. Thus a good record for sound construction management allied to a determined loss prevention programme will pay off in hard cash.

The Insurance Market

The degree of competition between rival insurance companies varies from country to country, and some have a single state-owned insurance company. Where competition does exist, the contractor is faced with a choice of how to place his business. There are clear advantages in avoiding confusion by covering the whole of one's insurance requirements with a single firm. However, many insurance companies themselves specialise in providing certain kinds of cover, so that a company which will provide a competitive quotation for 'contractor's all risks' may not be in a position to provide a performance bond.

Rates vary

In the insurance market, like any other market, the best way to buy is to emulate the wise housewife and 'shop around' until you find the most competitive quotation for each of your separate requirements. Even the large insurance companies and the firms linked to Lloyds of London, which write insurance policies in all categories, will offer the most competitive quotations in the areas they know best. Thus an underwriter specialising in building projects may offer a very competitive quotation for a contract to construct a school, but require much higher premiums for civil engineering jobs where he has less experience in assessing the potential risk. Taking this a step further, an underwriter interested in civil contracts may well quote a competitive premium for a road job but will only offer high rates for a contract involving irrigation, drainage or piling.

Using a broker

The contractor may reasonably object that he is too busy to spend time testing the market each time he needs insurance cover. Thus in countries where the insurance business is competitive there are advantages in seeking out a broker to act on your behalf. Most brokers have links with a wide variety of insurance companies and keep closely in touch with the full range of rates that are quoted for various classes of insurance, so they are in a much better position than the contractor to put together an insurance 'package' that is tailored to his needs and his pocket. Brokers obtain their income from percentage commissions paid to them by the companies with which they place the insurance, so they do not normally charge any fee for this service.

Know your broker!

Most brokers are honest professionals who will recommend the cheapest and best 'package', even if their commission earnings would be slightly higher if the policies were more expensive. They value the custom of their clients and take the long view, like wise contractors who seek a good reputation and plenty of repeat business. However, there are unscrupulous businessmen in every trade and profession (including contracting!), so the choice of a broker should be made with care and after consulting colleagues and business acquaintances (and perhaps your bank manager). Once he has chosen a good broker, the contractor will have the advantage of obtaining a competitive insurance 'package' specially tailored to his personal needs while dealing with a single office rather than trailing round from company to company.

Use your broker!

As in dealings with your bank manager (and for that matter with your family doctor!), it pays to be completely open and honest with your insurance broker. It is pointless to gloss over or hide risks that you know exist or are likely to occur, since their non-disclosure could invalidate the policy. Once he has a full understanding of the risks that are to be guarded against, the broker will be in a position to ensure that these are fully covered. The detailed policy wordings can differ in small but significant ways, and some companies might be prepared to cover your special risk for a standard premium while others would require an additional payment for cover that exceeded their own particular standard. Finally, should you be so unfortunate as to have to make a claim, the broker will be able to assist in filling in the claim form and advising on negotiations with the insurance company. In making a claim, never forget to tell 'the whole truth, and nothing but the truth'. To do otherwise would be to commit a criminal offence.

Chapter Nine

The Builder and the Law

The Legal Framework
The laws of a country provide a legal framework for all the actions of its citizens, although laws affecting contracts and business transactions vary considerably from country to country. Every action taken (and some actions that fail to be taken) and every transaction between two individuals or groups takes place within this legal framework and there is always the possibility of it giving rise to some form of legal action in the courts.

Avoiding Litigation
It is often said that the only people who do well out of the law are the lawyers. This is not entirely true, for the law is there to protect the innocent and to give protection and the chance of compensation to those who feel they have been wronged. However litigation is always costly and time-consuming, with the possibility that the opposing party might succeed, so the builder would be wise to do all he can to try to avoid it becoming necessary. The best way to avoid litigation is to know as much as possible of your rights and duties under the law, and avoid careless decisions or actions that could lead to trouble.

Negotiation
If a dispute does arise, it is always best to try to resolve it initially by negotiation. It is not always wise to stand too firmly on one's rights under contract law, and it does help if one at least tries to understand the point of view of the opposing party.

An Example
An example is the story of the punter who made a $1 accumulator bet on the outcome of a horse race. He chose a very unlikely combination of complete outsiders with the result that the combined odds, when multiplied together, were very high indeed. The horses all won, and his winnings came to $50,000. Unfortunately the bookmaker was a small

man (as bookmakers go) and said he could not pay the full sum immediately, since he would have to sell some property to get the money together. As a compromise he offered to pay $10,000 per week for five weeks. But the punter said 'No. I know my rights under the law. The contract between us is void. I want my $1 stake money back!' In real life no one is likely to be quite as silly as that, but the story does make the point that a little flexibility and readiness to compromise can lead to the most satisfactory solution.

Do It Yourself?

The law is not a suitable field for the 'do it yourself' enthusiast. The law is a very complex field of study, and the right course of action can only be decided on by a specialist who is aware of all the statutes and precedents. But every contractor should at least understand the general implications of a contract document before he signs it, should be aware of his responsibilities as an employer, including health and safety legislation, and should know enough general law to avoid the obvious pitfalls. If he operates his firm as a partnership or a limited company, he will also need to fully understand the special obligations imposed on these forms of enterprise. But if he receives a writ or a private client defaults on a certificated payment, he should consult a qualified lawyer. The fee will be a proper and legitimate business expense.

Law of Contract

Although the very act of setting up a business could involve certain legal liabilities, the contractor first really becomes aware of his legal rights and liabilities when he commits himself to his first contract to carry out specified work for a client.

What is a Contract?

A contract is an agreement between two or more parties which is intended by them to be legally binding. On large jobs it is standard practice for the contract to be based on formal contract documents prepared by the consultants, and the contractor is required to make his offer to build or 'tender' on the basis of those documents. On small jobs the contract may be based on a quotation supplied by the contractor, which effectively becomes evidence of the existence of a contract and its terms and conditions once it has been accepted by the client. A verbal agreement to carry out work in return for payment may even constitute a contract, but verbal agreements are very dangerous for a contractor since he will have no evidence if the client later goes back on his word.

Offer and Acceptance

A contract of any kind between two parties can only be lawful if they both agree to it at the time that it is made. But once they are both committed to it, either party can hold the other to its terms and conditions or seek damages as an alternative. Thus the time to check up on whether the contract will be advantageous is *before* you have either made a formal offer or accepted the offer of another party. The idea of *offer* and *acceptance* is crucial to the law of contract. To make a good contract one party must make a definite offer, such as a builder quoting a specified method of payment to construct a building on a particular site as shown on prepared drawings. But the contract only takes effect when the offer is subject to an unqualified acceptance by the client. If these two fundamental requirements are not met, the contractor may find himself in the law courts if things go wrong.

The Offer

A tender or quotation provided for a prospective client is an *offer* to enter into a contract, and it is open for acceptance unless it *lapses* or is *withdrawn*. There are three ways in which an offer may *lapse* automatically and cease to be available for acceptance:

1. If the contractor or client dies;
2. If the offer contains a time limit for acceptance and this is exceeded;
3. If there is no time limit, but acceptance does not come within a *reasonable* time.

Unfortunately the legal view of what is *reasonable* will vary from contract to contract, and it may be expensive to find out in the law court if a client tries to make you work to a fixed price tender submitted when materials and labour costs were much lower. Thus the wise contractor always includes a *stated time* for acceptance in his offer.

Withdrawing an Offer

What should the contractor do if he finds a mistake in his offer and wants to withdraw it? He must act quickly, and notify the client that the offer is cancelled *before* it is accepted. Once the offer has been accepted, it can only be withdrawn by *mutual* consent. In most cases the signing of the formal contract merely clarifies the position, and the legal contractual position dates from the *acceptance* of the tender. This is another good reason for checking — and double-checking — tenders before they are sent off.

Acceptance

For acceptance of an offer to be legally binding it must be clear and unqualified. A letter indicating an intention to accept, subject to certain conditions, is not legally binding. An 'acceptance' subject to conditions is regarded in the eyes of the law as a *counter-offer* rather than a true acceptance, and cannot give rise to a contract until the maker of the original offer accepts the new conditions.

Validity of Contracts

Most contracts work out reasonably well. The client gets his building. The contractor gets paid for it and — providing he knows his job — makes a profit. But the experienced contractor will know that disputes can arise, and then the validity of the original contract can come into question. Of course a contractor might like to get out of a contract if he is losing money, but that is not a valid legal reason — and the client could sue for breach of contract. There are seven main legal considerations affecting the validity of a contract:

Consideration: Under English or similar legal systems, a contract is an agreement to exchange something for something (e.g. a completed building for a monetary payment). No one can be required to do something for nothing;

Legality: If a contract is for a purpose that would break the law it is automatically void;

Intention: The parties must genuinely intend to enter into a legally-binding agreement;

Capacity: The parties must be legally capable of entering into an agreement, i.e. adult, sane and sober at the time (although drunkenness is not a very good defence!).

Influence: Consent must not have been obtained by fraudulent means or under duress;

Mistake: As previously stated, a simple mistake in calculating the estimate does not affect the contract. But if there is a major misapprehension about, for example, the identity of the other party to the contract, the contract might not hold good — but again finding out will probably prove expensive!

Misrepresentation: This can occur where an untrue statement by one party to a contract leads the other party to enter into it. This may be 'innocent' where the statement is honestly believed to be true, or 'fraudulent' where it is a deliberate lie. Either way it could endanger the legality of the contract, and a contractor should be very careful in ensuring that only factual claims are made about past experience, skills possessed and plant and equipment owned.

Terminating the Contract

Once the contract has been validly entered into, there are just four ways in which it can be brought to an end:

1. Performance
2. Agreement to Terminate
3. Frustration
4. Breach of Contract.

Performance

The most satisfactory way to bring the contract to an end is for both parties to do what they are required to do and thereby complete the contract to their mutual satisfaction. Thus the builder completes the structure according to the drawings and specification and within the contract period, the client meets the interim and final certificates, they shake hands and hope to do business again. Despite the warnings in this chapter, this still often happens! Good management should make any other outcome exceedingly rare.

Agreement to Terminate

It is always possible to bring a contract to an end by mutual agreement between the contractor and the client. Some contracts contain a clause that the contract shall automatically terminate after a certain time or if a certain event occurs, such as war or civil disturbance.

Frustration

Of course most contractors feel frustrated at one time or another — but frustration in the legal sense has a rather special meaning! It occurs where it is impossible to carry out the original contract due to a change in the law or a fundamental change in circumstances which could not have been foreseen when the parties entered into the agreement. It is however difficult to prove, and contractors should note than an unexpected occurrence that just makes the job more difficult or more expensive to execute will not be regarded as 'fundamental'.

Breach of Contract

If one of the parties to the contract neglects or refuses to meet its terms and conditions, this is a 'breach of contract' and the aggrieved party has a right of legal action against the other. There are two possible remedies against breach of contract:

1. Specific Performance
2. Damages.

Specific Performance

This occurs where the Court requires the party who broke his promise to actually carry it out. This is, however, not the usual remedy where a builder fails to complete the works set out in the contract. The more usual penalty is a financial one to offset the 'damages' suffered by the client due to the failure of the builder to deliver the right product at the right time.

Damages

The idea behind 'damages' is that the defaulting party should pay over a sum of money designed to put the aggrieved party in the position that he would have been in had the contract not been broken. If the job is abandoned when only half-finished, damages will be related to the costs of obtaining a new contractor to finish the work over and above what would have had to be paid to the original contractor. If a contract overruns (or will overrun due to abandonment by the original contractor), *liquidated damages* will also be assessed. These will be related to the losses suffered due to late delivery of the building to the owner — for example, rental payments on alternative accommodation. In general, damages will only be those reasonable and foreseeable as a result of the breach — but they can still prove very expensive for a defaulting contractor.

Damages Claims by Contractors

There are occasions where damages claims can work the other way and the contractor has a valid claim against the client. If the client will not allow the contractor access to the site, for example, he will obviously not be able to carry out the work. The same thing applies if his men are sent off the site for no good reason. In these circumstances it is likely that the contractor will have a valid claim for recovery of costs on the work he has done and also for the loss sustained due to being unable to complete the work (including loss of profit). But the contractor must understand that claims of this kind are hard to sustain, and if he just walks off the site with his men after an argument it is more likely that he will be found to be in the wrong.

Sub-Contractors

The contractor will be held responsible for failures and omissions of sub-contractors, even where these have been nominated or selected by the employer. The only exception is where the specialists are employed *direct* by the employer, in which case the contractor will have a valid claim if their actions disrupt his work. In general, a client will not sue a sub-contractor. He will proceed against the main contractor, who is responsible for co-ordinating all the work on the site. The only remedy that the main contractor can seek is by himself proceeding against the sub-contractor. This of course leaves the main contractor dangerously, and sometimes unfairly, exposed — since there is the risk that the client may win his case against the main contractor while the sub-contractor successfully defends his case. This is yet another reason for contractors to be very clear about their potential liabilities *before* they become committed to a contract.

The Client's Representative

In most building and civil engineering contracts, the client is represented by a professionally qualified architect or engineer — who undertakes the responsibility for designing and supervising the works. Actions or omissions by the client's representative can obviously greatly affect the cost and progress of the work that the contractor is required to carry out. The position of the client's representative is a difficult one, since most contracts allocate him a quasi-judicial role as arbitrator between the interests of the client and the contractor in addition to his direct responsibility as agent to the client — who is of course also his paymaster.

Legal Links

The legal links that are operative in most building contracts are illustrated in the next diagram. The main contractor is the focus of all relationships between sub-contractors and the client — and he cannot escape responsibility for the actions of his sub-contractors. The architect (or engineer) stands aside from these links, and may be called upon to decide which side is in the right in the event of a dispute. When measuring work for interim or final certificates, for example, the consultant (perhaps a specialist quantity surveyor) ought to ensure that the outcome is fair to the contractor as well as the client. In practice, most contractors try to get a little more than they are entitled to and a shrewd and well-organised contractor will claim fully for everything he might be entitled to. But the author has encountered instances of items being omitted, and in these cases it is quite in order for the omission to be drawn to the attention of the contractor.

The Consultant as Culprit

The contractor seeking to pursue a claim for a contract extension and the reimbursement of additional costs will have to prove that some other part or external circumstances caused the delay. Sometimes he will see the client's representative as the culprit. For example, if — despite requests from the contractor — the consultant fails to provide working drawings in good time, then the contract is likely to be delayed. If labour and plant on the site is made idle as a result, the contractor should have good grounds for a claim. Thus the client will have to pay, and the consultant will need to think up a plausible explanation to attempt to satisfy his client. Normally a contractor would be wise to avoid embarrassing the consultant for obvious reasons. If, however, there was no alternative a claim would be likely to be effective if it was proven that neglect occurred *and* that the consultant was acting in his capacity as agent of the client. If the consultant was acting at the time in his quasi-judicial capacity, it would be doubtful whether the client could be made to make financial restitution. The best approach for a contractor is to make every effort to avoid such a situation arising by working closely with the client's representative, and letting him know in good time before information and working drawings are required as well as pointing out errors and discrepancies as soon as they are noticed.

Take It or Leave It

In many cases the contract documents are prepared on the client's behalf by his professional representative. Although it may be open to a contractor to offer a conditional tender which would require specific modifications, most contracts are only available on a 'take it or leave it' basis. This does not mean that the contractor does not need to worry about how contract documents are worded. There are occasions where the wording is so favourable to the client (and unfavourable to the contractor) that the 'leave it' option is the right answer, and the invitation to tender should be politely declined.

The Risk/Reward Ratio

All contracts involve risk. Apart from making use of his managerial and technical expertise, the main reason for the client employing a contractor rather than doing the work by direct labour or force account is simply to pass the risk on to someone else. That 'someone else' is the contractor and, once his tender has been accepted, he will be the risk-taker according to the terms and conditions of the contract to which he is committed. His reward for carrying this risk is the profit which he expects over and above his estimated costs. The higher the anticipated risk, the higher will be the percentage addition for profit which he will include in his tender prices. If the ratio of expected risk to possible reward is unacceptably high, the intelligent contractor will keep clear of the project, and leave it to his competitors to cope with the problems and bear the loss.

Quotations

Some building contracts, particularly smaller jobs for private clients, are based on a quotation submitted by the contractor. In these cases the contractor obviously has much more scope to limit and define his risk by writing his own conditions of contract. They will have to be reasonably fair to the client or he will turn to another contractor instead. But the quotation and the conditions attaching to it must be carefully worded to ensure that the contractor is covered against all likely contingencies and, if he fails to do this, he will have only himself to blame.

Standard Contracts

It is helpful if contracts can be based on a generally accepted set of standard conditions such as the Royal Institute of British Architects for building contracts or the standard contract for civil engineering works issued by the Institution of Civil Engineers. These standard conditions can of course be modified or replaced by specific conditions drafted to suit the needs of any particular contract.

Twenty Questions

There are certain basic items of information and conditions that need to be laid down in every contract, so that they are understood and accepted by both parties at the outset. The following 'twenty questions' should be answered somewhere in the contract document, and the answers should be clear and unambiguous to avoid the risk of dispute or litigation.

The Contract 'Twenty Questions'

1. Name and address of client or employer?
2. Name and address of contractor?
3. Name and address of client's representative?
4. List of, and access to, drawings and contract documents? Address of site?
5. The contract sum and breakdown (e.g. by Bill of Quantities)?
6. Arrangements for interim and final payments?
7. Starting date and contract period? Duration of offer?
8. Provision for liquidated damages?
9. Relationship between parties and financial provisions if contract is abandoned?
10. Price fluctuations?
11. Provision for extras and variations from contract drawings and specification?
12. Prime cost sums and provisional sums?
13. Client's possession of materials and method of payment?
14. Access to the site for supervision of the work?
15. Responsibility for provision of tools and equipment?
16. Condemnation of unsatisfactory materials and workmanship?
17. Responsibility for fees payable to water authority, local authority, etc?
18. Compliance with site safety and labour legislation?
19. Insurances and indemnity from sub-contractors' claims?
20. Agreement to methods of arbitration in the event of a dispute?

1. Identity of Client

The contractor needs to know who (or what organisation)

is his ultimate client. If the client is a person, then the contract could be voidable if the individual concerned is of unsound mind, an infant or otherwise not regarded by the law as able to execute a contract. If an organisation — private or public — is the client, the person who signs on its behalf should be checked as being authorised to do so. If the worst comes to the worst and the client fails to make proper payment for work done, an action in the courts may fail if the client is not correctly identified.

2. Identity of Contractor

Equally the contractor must be correctly identified. Some contractors occasionally operate through partnerships and joint venture operations or undertake specialist work through subsidiary companies. A misunderstanding as to the nature of the contracting party might be an additional source of difficulty if a dispute arises.

3. The Client's Representative

The third party who must be identified is the architect or engineer who will act on behalf of the employer as his professional adviser, agent and representative. Most of the contractor's correspondence, discussion and negotiation will take place with this consultant, and he will need to know who he is dealing with. Apart from legal requirements, the experienced contractor will come to know which consultants he can work with efficiently and amicably. He is likely to quote keener prices on invitations to tender from these firms, since he will know that the margin of risk will be lower.

4. Site Address. Drawings and Documents

The address of the site may sound too obvious to be worthy of mention. But the author has encountered instances of the right building being constructed at the wrong place. One example was a contractor who made a successful bid to build a headmaster's house at a school some way from his home town. The contractor failed to arrange a site meeting with the clerk-of-works to agree setting out, and went straight to what he thought was the right school and saw the headmaster. The headmaster did not show his pleasure and surprise at the contractor's appearance, and merely showed him where to start digging the foundations. After a few weeks the headmaster of the school where the house *should* have been built complained to the Ministry of Education that a start had not been made, the chain of complaint continued through the Ministry of Works to the clerk-of-works, who wrote a severe warning to the contractor that he risked withdrawal of his contract. The outcome was that the unfortunate contractor faced a dead loss on the labour and materials put into building the house in the wrong place, upset his client, ran the risks of liquidated damages on his real contract — and became something of a laughing stock among the local community!

Besides setting down the site address, it will be helpful to identify all applicable contract drawings, documents and specifications upon which the bid is based. This is vital if there have been amendments to the contract drawings and stated requirements after the contract documents have been received from the client's representative. For example, the client may have made a late decision to omit one of a series of buildings, so the contractor has lowered his price accordingly. If this omission is not clearly set down in the bid or quotation, the contractor could find himself liable to carry out work which he did not allow for in his revised estimate.

5. The Contract Sum

The contractor must make sure that his calculations leading to his bid figure are both checked and double-checked. It is very easy to make a mistake or omission in totalling the separate bills of quantities or adding up the separate figures to produce the grand total. Errors could invalidate the tender or, in some circumstances with a ruthless prospective client, lead to the contractor being legally committed to carry out work at a certain loss.

It should be standard practice for the contractor to set down his bid *in writing* as well as in figures in the same way as a cheque is written — to ensure that the figure cannot be misunderstood or tampered with by any interested party.

Where the contractor is providing his quotation for a private client and is free to set it out as he chooses, the bid price should appear near the top of the proposal. If a great many rival bids are received, the buyer may look at them very rapidly and only choose the most favourable for closer examination. If he cannot see the offer price at first glance, he may discard that quotation with the result that all the estimating work is wasted.

Unless this is covered in the invitation to tender or the contractor is prepared to offer a discount for prompt payment, it is wise to state firmly as a condition of the bid that no discount will be deductible. A suitable phrase is:

'All prices are strictly nett and no discounts will be given unless otherwise stated.'

In an invitation to tender based upon a bill of quantities, the contractor should be sure that every item is priced. If there are any items for which he does not wish to charge separately, such as providing a brush finish to concrete paths, this should be indicated by writing 'INCL'. (short for 'inclusive') in the unit price column. It will then be clear that the contractor is covering this cost in his unit price for laying the concrete.

6. Arrangements for Interim and Final Payments

The terms of payment are a common cause of dispute between contractors and their clients. For his own protection the contractor must ensure that these are quite clear before he puts his name to any bid or quotation. Most contractors possess only limited working capital, and rely on prompt interim payments to finance work in progress. In the absence of a specific condition covering interim payments, the client could argue that no payment need be made until the work is complete. Most standard contracts provide for interim valuations and payments, subject to a percentage deduction for retention money as a protection for the client. It is common for the retention deduction to be 10 per cent on interim certificates and 5 per cent on the final account, which will be released to the contractor when any necessary remedial works have been carried out at the end of the maintenance period. The *duration* of the maintenance period should be stated.

In many standard contracts the contractor is required to complete the work even if the client is in arrears in honouring interim certificates. There is not much the contractor can do about this except check on the reputation of the prospective client for the promptness or otherwise of his payments, and decide whether the risk is so great that it would not be worth submitting a tender.

If the contractor is submitting a quotation for work, he will be able to cover himself with a clause such as:

'TERMS OF PAYMENT: Monthly interim payments shall be made as the work proceeds of 90 per cent of the work executed and of materials on the site (whether fixed or unfixed), such payment to be made within fourteen days of application. If default is made in any payment due then the contractor may suspend or abandon the work and remove unfixed materials, tools and other equipment from the site. Five per cent shall be released to the contractor immediately after issue of a certificate of practical completion (which shall not be unreasonably withheld) and the balance six months thereafter.'

It is very much in the contractor's interest to obtain a certificate of practical completion as soon as possible. Not only does this remove the shadow of liquidated damages and result in the release of half of the percentage retention, but also the defects liability period (or 'maintenance period') starts on the day that it is issued. In the RIBA form of contract, the schedule of defects has to be delivered to the contractor by the employer's representative within 14 days of the expiry of the defects liability period. Technically it would be possible for the contractor to reject the list if it arrived after that time, but it would be pointless — since the client could then sue him under common law for failing to carry out the work satisfactorily. The contractor should, however, remember that under the RIBA form of contract he will not be required to do any work other than that specifically listed on the Schedule of Defects. Once that has been issued, the contractor must complete the work listed — but he cannot be called upon to carry out any further work.

7. Starting Date and Contract Period. Duration of Offer

On contracts with a tight schedule for completion, the time limit for setting a starting date after acceptance of the tender will be of importance to the contractor. The contract is legally in being from the date that the client *accepts* the tender. It will normally be to the contractor's advantage to allow himself a reasonable time for ordering materials, planning and scheduling before physically commencing work on the site. Some contractors think it creates a good impression to start almost immediately, but this can lead to confusion and inefficiency — besides bringing forward the completion date and risking liquidated damages. The right answer is to set the starting date for as soon as the contractor can be reasonably sure of having all the necessary resources of men, materials and equipment available to ensure continuous and workmanlike progress.

The usual practice is for the contract period to be stated as a part of the invitation to tender, but there are occasions when the contractor is asked to submit his own completion time with his financial quotation. In the latter case competing tenders will be evaluated on a time *and* cost basis, and the lowest tender may be rejected in favour of an offer of more rapid completion. In these circumstances, the contractor may offer alternative prices and completion times so that the client can choose whether it would be worth spending more money to get the building more quickly. In such a contract, by implication 'time is of the essence' and a contractor who overruns will be in a weaker position in defending a claim for liquidated damages.

When a contractor tenders for work he will not want the tender to remain open for acceptance for an unlimited period. Prices and wages can rise very rapidly and endanger the profitability of the contract. If there is no specific condition limiting the duration of the offer, the legal view would be that it should be allowed to lapse after a 'reasonable' period. But finding out what would be regarded as reasonable in any particular case would be expensive in a court of law. It would be much better for the contractor to protect himself by inserting a clause such as:

'PERIOD OF TENDER: This tender is conditional upon receipt of acceptance by us in writing within thirty days of the date hereof or such longer period as may be agreed.'

8. Liquidated Damages

If a contractor overruns his contract period he will be technically in breach of contract — even if he goes on to complete on the following day or in the following week. Thus the client is entitled to receive a sum of money in compensation for not having the use of the building on the date promised. This sum of money is normally stated in the contract and can either be in the form of *Liquidated Damages* or a *Penalty Clause.*

Liquidated damages are stated as a sum payable for every day of delay after the agreed completion date, and the actual amount will be the unit sum multiplied by the number of days. It should be made clear whether 'days' means working days only or includes holidays. The general principle behind liquidated damages is that they should be related to the actual loss suffered by the client through not having the promised facilities available on the contract completion date. Thus they are a *measure* of the likely loss that might be suffered, e.g. loss of revenue from rents on houses. If it is based on a genuine *pre-estimate* of the loss, it will be allowable in full. If no clause is included in the contract, it is possible to sue for *actual costs* incurred.

A penalty clause, on the other hand, may be quite unrelated to any calculation of losses suffered. It is there, like a fine or a prison sentence in criminal law, as a simple deterrent to the wrongdoer. If the contractor originally agreed to the inclusion of such a clause in the contract, it will be no defence to suggest that the sum exacted is extravagantly high. The risks and effects of transgressing a penalty clause must be carefully evaluated by the contractor *before* he submits his tender. One comfort for the contractor is that penalty clauses are rather difficult to enforce — but it is still a risk that is better avoided. Most contracts based on standard conditions provide for the possibility of extensions to the contract period to cover unforeseen delays of various kinds, particularly those resulting from a mistake or action by the client or his agent. Any such delays should be carefully recorded in the site diary, and covered by a letter to the consultant requesting an extension. Even if the contractor is confident that he will be able to catch up on the time lost, it is best to be on the safe side.

9. Abandonment of the Contract

Once the tender is accepted and a contract exists, there is always the possibility that it might be abandoned for some reason. Unless this is done by mutual agreement, it will mean that one of the parties will consequently be in breach of contract and the other will be in a position to claim damages.

If the contractor is at fault, the ultimate sanction is for him to be expelled from the contract and most standard conditions contain provisions for this. For example, under clause 63(i) of the Institution of Civil Engineers Conditions of Contract (fourth edition) the client may expel the contractor if he abandons the contract, suspends progress or fails to proceed with 'due diligence'. The general procedure is that the client's representative first issues a *Notice of Default* and is not then obliged to pay any further money to the contractor until overall completion of the project. The client, if the contractor is expelled, is entitled to claim from the contractor the additional costs resulting from having the work completed by another firm, and may be able to seize as security any plant, materials or temporary works owned by the contractor.

The eventual damages that would be payable in the event of a contractor abandoning his contract would have to be decided by agreement, negotiation, arbitration or in the law courts. The actual sum would be made up of two parts:

1. The difference between the price of the work as agreed in the contract and the actual cost to the client of its completion;

and

2. The loss of rent or value of occupation in consequence of the delayed completion.

If it is the client that is at fault, it is the contractor who will be entitled to damages for breach of contract. If the employer gives notice to the contractor not to do any more work, this amounts to a total breach of the contract — although a notice to postpone work temporarily would not have the same effect. Normally, provided he could prove that it was his client who was in breach of contract, the contractor could claim for all work done and materials supplied, consequential losses on leaving the site plus the loss of profit that would have been enjoyed if the contract had proceeded. However, the assessed damages will be what the arbitrator or the court regard as 'fair and reasonable' and they will take note of any contrary or mitigating points put by the employer.

If the contractor is putting forward his own quotation, a useful safeguard would be to include the condition:

'SUSPENSION OF WORK: If work is suspended, all consequential costs will be borne by the client. If work is later resumed the costs of cessation and resumption should be the responsibility of the client.'

It will be remembered that an earlier suggested condition would give the contractor the right to suspend the work if the client defaulted in making payment.

10. Price Fluctuations

When a country is undergoing severe price inflation, contractors working to fixed price contracts can suffer grievously. Thus some form of condition to permit reimbursement of price increases affecting labour wages, materials and plant costs should be included as a condition of contract if at all possible. Unfortunately many contracts are only open for tender on a fixed price basis. In these cases the contractor needs to build an allowance for likely price inflation into his tender. By doing so, he of course risks losing the job to a less cautious contractor, but there is no point in working at a loss.

11. Extras and variations

Most contracts contain provisions for the amendment of the detailed design to suit changed circumstances or an alteration in the client's requirements. Thus 'working drawings' may differ from 'contract drawings' and 'A', 'B', 'C' etc. revisions may be issued as the work proceeds. On one contract the author recalls several drawings that went right through the alphabet, so the designers went on to revisions 'AA', 'BB', 'CC' etc. If the revisions come in time, they will be an annoyance to the contractor but should not cost him any money. But if they come after work has started on the previous details, abortive work may have to be demolished and additional work will have to be done that will not be covered by a measurement at bill rates.

Thus the contractor needs to make certain that he will be able to recover all the additional costs, including overheads, that will be associated with extras and variations. Any such alterations should be covered by written variation orders, signed by the architect or clerk-of-works, as they occur. Documentary cost evidence, such as time sheets and plant and material returns, should also be agreed and signed by the client's representative *as the work proceeds.* Where a great deal of money is at stake, it may even pay to take photographs as evidence of site conditions in substantiation of a claim.

12. Prime cost sums and provisional sums

Provisional sums and prime cost sums are both amounts of money set aside in the Bill of Quantities for expenditure at the discretion of the client. These amounts will only be spent at the specific written request of the client or his representative.

An item covered by a provisional sum may be executed by the contractor's own staff or by a nominated sub-contractor. The client pays for the cost of the work, including overheads and profit, either on daywork rates or on lump sum or measured rates as agreed when the work is authorised. The usual reason for including provisional sums is that the designer has inadequate knowledge of the site conditions or his client's requirements at the stage of going out to contract, and needs to keep his options open until these uncertainties are clarified.

Prime cost sums usually refer to items ordered and paid for by the client, but which the contractor will be required to incorporate in the building. The contractor is usually paid for handling, storing and building in the item at either daywork or agreed rates.

13. Client's possession of materials and method of payment

Most contracts provide for all materials on the site to be deemed to be owned by the client as soon as they are delivered. This can cause problems for the innocent supplier of a contractor who goes bankrupt in the middle of a job, as the supplier will be unable to secure payment or even take his materials back from the site. Most contract conditions do allow a contractor to include for 'materials on site' when applying for an interim certificate, although only a proportion of the full value is payable on perishable items such as sand, aggregates and cement. Naturally the client will make no payment in respect of materials that are surplus or do not conform to the specifications.

14. Access to the site for supervision of the work

There are usually specific provisions that the contractor should give adequate notice at various stages of the work so that it can be inspected by the architect or clerk-of-works. This applies particularly to parts of the works that will be covered as the job proceeds, such as foundations, runs of drainage pipe and steel reinforcement that will be surrounded with concrete. A contractor who is foolish enough to carry on with the work, without allowing an opportunity for an inspection to be made, runs the risk of having to demolish or remove later work at considerable cost. Even if the original work is then proved to be perfectly satisfactory, the contractor will have to bear the cost because he did not conform with the provisions for supervision in the contract.

15. Responsibility for provision of tools and equipment

In most cases these are the sole responsibility of the contractor, but it is worth checking the documents to see if the client will be prepared to supply any specialist items. If such items can be supplied, the contractor should also check on whether the client will make a charge for their use so that he will be able to adjust his estimate accordingly.

16. Unsatisfactory materials and workmanship

There are provisions in all contract documents for the architect, engineer or clerk-of-works to condemn materials, components or parts of the structure that do not conform fully with the drawings and specification. The contractor should never attempt to substitute alternative items for those specified without obtaining prior permission in writing, even if the original item is in short supply and he honestly believes that the alternative will serve the purpose just as well.

Particular attention must be given to the quality of concrete, since cube tests only reveal inadequate strength characteristics well after the concrete has set, so remedial work can be very expensive. It is false economy to try to cheat the client by using less cement than is specified for the mix. It is also dangerous to introduce too much water into the mix to make it more workable, since there is a close link between strength characteristics and the water/cement ratio.

17. Responsibility for fees payable

Fees are sometimes charged for the provision of information or building inspections by the local authority, or the protection or diversion of water pipes or electricity service cables. The contractor should examine the documents carefully to ensure that all fees will be reimbursed by the client. If this is not the case, he will have to recover them in some way from either his estimate bill rates or his overheads charge.

18. Compliance with site safety and labour legislation

There can be no question that the contractor should comply with the law of the land, but the contract documents may make additional demands regarding safety procedures, employment procedures and welfare facilities. Compliance with these demands can be expensive, and it is therefore important that they should be evaluated and costed at the estimating stage so that they are reflected in the contractor's tender price.

19. Insurances and indemnity from sub-contractors' claims

The complex interlocking legal relationship between the client, the consultant, the main contractor and his sub-contractors (both nominated and direct) can easily give rise to problems if anything goes wrong. Thus the clauses that describe this relationship and define the precise responsibilities of the contractor must be examined with special care. If the conditions seem unusually onerous, consult your legal adviser *before signing your name to the tender.* Once the contract is binding it will be too late for regrets. If the risks are great but the contract is otherwise attractive, it may be possible to arrange some form of insurance cover. Alternatively the contractor might make a conditional bid that would only be open for acceptance if the burdensome conditions were excluded or modified. However, a conditional bid should be seen only as a last resort, since such bids are unpopular with both clients and consultants, and some make a practice of rejecting such bids on principle.

20. Arbitration

Naturally we hope that no dispute will arise during the course of the contract. But differences of opinion and interpretation do occur, and settling them in a court of law can be both time-consuming and very expensive. Much can be achieved in a friendly spirit of 'give and take' with the client's representative, but it is not always possible to resolve arguments — particularly those involving substantial claims — in this way. In such a case, both parties would save money by submitting the dispute for arbitration by an independent person who would attempt to achieve a settlement without the need to go to court. Thus it is helpful if the contract documents contain a specific provision for arbitration in certain circumstances.

Chapter Ten

The Contractor and the Client

Pleasing the Client

The first chapter of this book introduced the idea of marketing as a way of obtaining clients. This final chapter takes as its theme the equally important question of how to work with the client and his professional representative, how to please him and how to keep him. Clients are hard to find and easy to lose, and a contractor who finishes a contract on time and with a good reputation will be one step ahead of his competitors when the next invitation to tender comes along.

The Site

As far as the client is concerned, what happens on the site is all important. He won't really care whether the contractor has an office or not — few clients visit contractors at their offices. But the client — and his architect — will see the site, and see it often. After all, the client owns the site and the contract has been awarded to build *his* building to suit *his* requirements and will be paid for with *his* money; so it is not very surprising that he will want to feel pride in his possession as the structure gets underway. It should be noted that this 'pride of the proprietor' is exhibited just as much by the senior officials of a public client, and is quite legitimate since they will have devoted much time and effort in getting the project to the construction stage.

Site Organisation

A tidy and workmanlike site is important. First because it helps the contractor to operate more efficiently, and saves wasted time and wasted resources. Secondly, because it shows the client that the contractor takes pride in his work and is keen to give a good impression. The client is likely to be an amateur in the building business, and even his architect may be much more expert about construction theory than construction practice. Thus they will tend to judge the quality of his management by superficial impressions, and a clean-looking site with everyone on it appearing to be working purposefully at a defined task will give the outsider an impression of a contractor who knows his job.

The Site Office

On all but the very smallest jobs, the site supervisor should be provided with an office where he can keep and study the drawings, specifications and other documents relating to the job, keep and update his bar chart programme and prepare labour, plant and material returns as required by the head office. It need not be elaborate, but it should have sufficient basic furniture and equipment to enable him to run the site on your behalf in a businesslike way. When you visit the site, make a point of looking round the site office to ensure that working drawings are up to date (with the latest revisions in place of superseded editions), there is a file of clerk-of-works' instructions and variation orders and all the other necessary information is readily available.

Train your Foreman

It is quite common for good practical foremen to find it difficult to keep site records efficiently. It is well worthwhile taking a little time to explain (patiently!) the reasons for requiring them to keep their site office tidy and submit time sheets, etc. regularly, rather than simply saying 'it has got to be done'. The foreman is in the front line of management, and he will almost always perform better if he can be helped to understand that he is a member of a team rather than an obeyer of orders. Show him how to *use* a bar chart to plan the day-to-day tasks for the operatives on his site and plan his material purchases in such a way that hold-ups are avoided.

Waiting Time

'Waiting time' is a far too common occurrence on many building sites, as the carpenter waits to be told what to do next, the plumber waits while the foreman goes off to the supplier for an urgent item and the concreting gang has to stop until a fitter comes out to repair the mixer. 'Waiting time' is a polite way of describing 'wasted time', and wasted time means wasted money. Can it be reduced or even eliminated? Usually the answer is yes — providing site management plans ahead.

Site Layout

Studies have shown that up to half the man hours expended on a building site are concerned with the movement and handling of materials. Much of this time could be saved,

leading to greater productivity and lower costs, if the site layout was planned from the start of the job to minimise unnecessary movements. The key factors in designing an efficient site layout are:

— Access roads
— Water and other services
— Materials stores and stockpiles
— Placing of plant and equipment.

Access Roads

If the construction of permanent roads is included in the contract, it may be sensible to undertake the drainage and basic road work (excluding the finishing course) at the start of the job to obviate the need to lay down and later remove a network of temporary access roads. Otherwise temporary roads should be levelled and a sufficient base course should be provided to cope with the traffic that is likely to use them. In some cases simple tracks for wheelbarrows will be sufficient, but even these should be made reasonably level and smooth so that they can be used without unnecessary effort and spillage of materials.

Water and Services

Concreting is a key operation in most construction activities, and concrete cannot be mixed without water. So a water service or supply should be arranged right at the start of the project. If the contract calls for the provision of a permanent water supply, this could be combined with the temporary supply to cut out the need for extra excavation and piping.

The best way to move water around the site is in a pipe. The worst — and most wasteful and expensive — way is to pay out wages for people to carry it around in buckets. So make the maximum use of pipes and hoses to get the water close to where it is going to be used. If an electricity supply is to be provided, it may be helpful to provide this early in the contract for lighting purposes or to power small tools such as drills and sanders.

Materials Stores and Stockpiles

The siting of store sheds and stockpiles should be carefully planned so that materials will be available as close as possible to where they are to be used, but also in such a way that they will not interfere with building operations or block accesses. Valuable items should be stored in locked sheds. Aggregate stockpiles should be sited as close as possible to the area where the concrete will be mixed. Make sure that there will always be good road access through the site to the storage area, so that materials and components can be unloaded directly from delivery trucks without unnecessary handling or transport. Remember 'materials on site' are the property of the client, so he has a right to expect that his possessions are stored and handled with due care and attention.

Placing of Plant and Equipment

An area should be set aside for the storage, maintenance and repair of mechanical plant and equipment, together with diesel tanks, etc. if these are to be provided on the site. If a hoist is to be employed, it is worth taking care to think about its usage at various stages of the construction work so that it can be sited at a point which will minimise material movements within the building. Concreting and blockmaking areas should be sited to minimise transport of the finished product, and it may pay to resite these areas at various stages of the work. Handling of wet concrete should be cut to an absolute minimum for reasons of quality control as well as economy, and finished blocks are both more delicate and more difficult to handle than their component materials.

Choice of Plant and Equipment

An important aspect of site organisation is the choice of appropriate tools, plant and equipment. Simple equipment is often the most productive. One man with a wheelbarrow can sometimes get through as much work as five with headpans. Well-made timber formwork can be used over and over again, providing it is kept clean and is carefully handled. Expensive mechanical plant is only worth buying if you have enough work to keep it fully employed, and if you have trained and motivated operators who will ensure that it is properly used and regularly maintained. Even the choice of simple hand tools, like spades, hoes and shovels, is worthy of time and consideration. You cannot expect high levels of productivity from men with poor tools, and it is well worth paying more for tools that are properly made for the job and are sufficiently robust to stand up to hard work. Calculated on an hourly basis over their useful life, these tools are much cheaper than the wages of the men who use them yet the introduction of appropriate tools can increase productivity by as much as 20 or 30 per cent.*

Planning Plant Requirements

On jobs where mechanical plant has to be used, it is worthwhile to produce a special programme to co-ordinate its use and ensure that it will be available when most needed. If two separate contracts will require a particular item of plant at the same time, priorities can thus be decided well in advance or arrangements can be made to hire plant to cover the peak period.

A PLANT PROGRAMME

JOB NO........ PLANT PROGRAMME										
ITEM	WEEK									
	1	2	3	4	5	6	7	8	9	10
Excavator	▓	▓								
Conc. Mxr.	▓	▓	▓	▓	▓	▓	▓			
Generator			▓	▓	▓					
Crane				▓	▓					
Hoist						▓	▓		▓	▓
Dumper	▓	▓	▓	▓	▓	▓				

Supervision

The site supervisor is the contractor's front line manager. He represents the contractor in day-to-day contacts with the firm's direct employees and sub-contractors and their performance will be directly affected by his skill or otherwise as a practical manager. He will also be regarded as the con-

*See **Better Tools for the Job,** Intermediate Technology Publications, 9 King Street, London WC2E 8HN.

tractor's representative by the architect or clerk-of-works, and he will have to be tactful and diplomatic in his dealings with them, since most contracts give very substantial powers to the client's representative. If a clerk-of-works feels like being really unfriendly to a contractor, he can apply 'the letter of the law' and reject materials and components that are adequate but not quite perfect, and require the demolition of completed work that is marginally substandard. By acting as the contractor's 'ambassador' and building up a friendly and mutually respectful relationship with the client's representative, a good agent or site foreman can ensure that problems are tackled in an understanding way — so that the client's interests are not neglected, but the contractor too does not lose out.

Helping the Site Supervisor

If he is to perform well, the site supervisor needs the confidence and help of the contractor himself. It is hopeless just to pick on the best craftsman on the site, give him a site office, put up his wages by a few cents an hour and hope that he will perform. To start with, he will have to be able to read drawings and specifications so that he will be able to describe how the work is to be done, and correct mistakes before it is too late. He will need to establish his authority on the site, and the contractor can help him by treating him with proper respect when he visits the site. If he has any orders to give to the site staff, he should tell the supervisor, who will pass on the message. Equally, complaints from site staff should only be dealt with after hearing the foreman's side of the story. Men quickly come to know when their foreman has lost the confidence of his boss and will deliberately by-pass him, leaving the supervisor with a kind of empty responsibility but without the authority with which to get results.

How many Foremen?

This depends mainly on the nature and the complexity of the work, but also of course upon the ability of the foreman himself. Generally, it is difficult for one man to supervise more than 20-25 others and be fully effective. Once a work-force exceeds this number, it is better to have a general foreman responsible for the whole site working through charge-hands for the main trades or working groups.

Communication

As a business grows in size, the owners and top managers can no longer rely on orders and information being passed around by word of mouth. The problem of *communication* arises, and they have to think about ways in which they can ensure that orders, opinions, ideas and information can be passed quickly and effectively through the organisation to the target person or group. It is not just a question of getting orders down from the boss to the man with a shovel on the site. That boss will only make good decisions if he has the right information at his fingertips, so he also needs a system that will ensure that he gets to know facts and ideas that start out further down the chain of command. Very often 'the man who knows most about how to dig a trench is the man with the shovel in his hand'. The wise contractor will listen — and make sure his foremen listen — to the manual employees, craftsmen and operators who undertake the physical work on the site. Not only will it improve their motivation by making them feel part of the team, but very often they will come up with suggestions for improved methods which will help them *and* help you.

Communication and Hierachy

Communication is tied up with the idea of hierachy, or the way in which responsibilities are shared out within the organisation. The simplest possible hierachy for a contracting firm is that shown below, in which there is a single employer who takes on a group of workers to carry out some task:

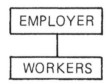

This usually applies only to very small firms and 'labour-only' gangs. Communications problems should hardly occur at all, since the boss will know all his employees and their strengths and weaknesses, will probably see them every day to check on their work and tell them what to do next and the workers, in turn, know that they take their orders direct from their boss and he is the man they have to satisfy.

When the Firm Grows

When the firm grows, so do the problems and a point arrives where the sole proprietor simply cannot continue to run the firm single-handed. Either he will take on a partner, and they will split up the work between them, or he will employ some kind of manager, who will take his orders from the employer but will be in a position to give orders to the other employees. The problems of growth are difficult to cope with, and many a good small firm has been crippled by trying to take on too much extra work too quickly. The problem is not just the need for extra resources — particularly money and manpower — because in addition the organisation of the firm itself has to be developed and adapted to cope with the growing responsibilities. In fact the complexity of the organisational problems can grow a lot faster than the firm's annual turnover, and it has been suggested that when a firm's turnover grows tenfold, the organisational problems that have to be tackled will grow by as much as a hundredfold!

Several Sites

By the time that the firm has reached the stage where several contracts are underway at the same time, the owner may have taken on a Contracts Manager to deal with the direct management of the sites, so that he can concentrate on commercial aspects and general development of the business. He may also have taken on a book-keeper to ensure that the records are properly kept, and perhaps come to an arrangement with an accountant to work with the firm on a part-time basis. On the larger sites, there will probably be a site agent, assisted by trade foremen and a site clerk. The smaller sites, and those close to the office, will probably be looked after by a working foreman. If this is how the business has developed, the organisation chart will now be as follows:

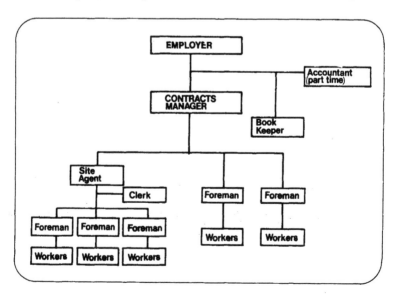

More Complicated

Already the firm has become a much more complicated organisation, not just for the people who work within it, but also for outsiders including the client, his representative, suppliers, the bank manager and all the others who have dealings with the firm. There are now five tiers in the organisation and, on the larger sites, three intermediate levels of management separate the employer from the manual workers on the site. This means that the chances of confusion and misunderstanding are also much increased, and it will need a determined effort by everyone from the employer downwards to ensure that orders, information, messages and information get to the right people and are acted on. This will only happen if the fundamental rules for successful organisation are followed by everyone concerned.

Rules of Organisation

An organisation is a grouping of several people, who get together to undertake a task that is too much for any of them to tackle on their own. To make it work, certain fundamental rules have to be observed:

1. The objectives must be clear so that everyone knows what has to be done;
2. The channels of command, as expressed in the organisation chart, must be clear. To put it simply, everyone in the firm must know who his immediate boss is and from whom he should seek instructions;
3. The status of groups of people within the organisation should be based on some clear division of responsibilities, related to the overall objectives and tasks of the firm as a whole;
4. Every individual should know exactly what is expected of him, preferably in the form of a written job description. Whatever responsibilities are allocated to him, it is vital that he should also be given the *authority* to carry them out;
5. There should be clear arrangements for monitoring and controls of the firm's activities, so that deviations from policies, plans and programmes can be promptly detected and corrected.

Delegation

Whatever the organisation pattern, the employer in a growing business has to learn one important skill — delegation. The advice 'don't keep a dog and bark yourself' is easy to give, and often easy to accept, but sometimes hard to apply. We all tend to be best at – and happiest performing – the jobs we are most used to. But the man who is running an expanding building firm finds that he has to spend more and more time on the commercial aspects of his business, and so has less and less time to direct technical operations on the sites. This is quite normal, because the paperwork involved in accounting, estimating and record-keeping for a growing firm is itself very considerable, and the systems of reporting and recording have to be set up and modified to ensure that they are fully effective. In addition there are some outside contacts, such as those with clients, important suppliers and the bank manager, that can only be undertaken by the principal of the firm.

Where are we going?

Then there is the absolutely vital – but often squeezed-out – time that should be devoted to sitting back and thinking carefully about where the firm is going, and how it is best to get there. The setting of general policy can only be done by top management. It is the one task that just cannot be delegated. If top management doesn't give time and effort to working out a corporate policy, the firm as a whole will soon lose its sense of direction. For all these reasons delegation is a must, and competent staff must be engaged and given the responsibility and authority to make day-to-day decisions.

Trust Your Staff

Building is a business where problems arise suddenly, and have to be solved quickly to avoid disruption and additional costs. This is another reason for *decentralising* routine decision-making to the lowest possible level in the organisation, by trusting your staff and *delegating* sufficient authority to them to get things done on your behalf.

Can they be trusted?

The immediate objection is 'they can't really be trusted, and if something goes wrong *I* shall be the one who pays the bill, not them'. But if your foreman cannot be trusted, they shouldn't have been employed in the first place. If they were all you could get for the wages you were prepared to pay, then maybe you should pay higher wages and take on people who are more competent. But the crucial answer is proper definition of the scope of individual jobs plus a system for reporting back and ultimate *control*, so that a mistaken decision at a lower level can be picked up quickly and corrected as necessary. If the front-line managers are given clear job descriptions *and* are helped to understand the policy and objectives of the firm, it is likely that they will respond to your confidence in them by trying hard to do what you would have done in their position.

Training

A further aspect which characterises the really successful growing firm is a positive attitude towards training at all levels of the organisation. There is little point in the employer going off on a training course and becoming an expert in the techniques of critical path programming, if his foreman can't understand a bar chart and the concreting gang splash extra buckets of water into the concrete mix to make it more 'workable' because no one has ever told them this will seriously weaken the finished concrete.

Not just 'Courses'

Training doesn't just mean sending people off on courses. Some of the best practical training can be done 'on the job'. Any good manager should see himself as a part-time training officer for his subordinates. In the above example, the employer might bring in his contracts manager when he was discussing the balance sheet and profit and loss account with his part-time accountant. The contracts manager might take a little time to explain how to use and update the bar chart programme on the site of a newly-promoted foreman. The good foreman, in turn, will not just give orders, but also explain the reason why.

Will they stay?

Some contractors are afraid of training their employees, because they might leave and set up in opposition or at least ask for higher wages or salaries. In truth, ambitious people are more likely to leave because they feel that they are not learning anything at work than because they are improving their capacity and performance. If the business is expanding, it is likely that there will be opportunities to make internal promotions so they will have the chance to earn more as they become more valuable members of the team. 'Home grown' managers are usually the best for the health of the organisation, because they take less time to get acclimatised to the firm and its approach to the contracting business. They also start with the advantage of knowing the strengths and weaknesses of your existing staff. Another advantage of an emphasis on training and internal promotions is that it is good for the morale of the remaining staff, who will see that good performance is going to be appreciated and rewarded.

Flexibility

It is not possible for the management expert to prescribe some ideal form of organisation for a building firm of a certain size, as a doctor might suggest a bottle of aspirin tablets for a man with a bad cold. To take the medical analogy a stage further, the first step for the 'management doctor' is to diagnose the illness. Only then can he write down an appropriate prescription, in the form of a management chart that will fit the needs of that particular business and the work style and experience of the people at the top of the firm already.

Keep it Simple

One good principle is 'keep it simple'. The author has encountered firms where the organisation chart has been as complicated as an oriental carpet and, if the employees had really tried to run the firm 'by the book', it would have rapidly come to a complete standstill. The anti-organisation man might say 'that proves that organisation charts are a waste of time anyway'. But that is not quite fair, because even the inefficient organisation has a system of some kind; the trouble is that it is not written down and has come about by accident to suit the preferences and eccentricities of individuals rather than the needs of the organisation as a whole. If the site foreman of a contractor with several thousand employees walks straight into the Chairman's office with an order for a few bags of cement, this is one way of running a business and the contact between them can be simply described by a line on an organisation chart. But it is still a crazy way of running a business!

Division of Work

In deciding on the most appropriate form of organisation, there are three obvious ways of dividing it up for a typical contracting firm:

1. By product or type of service
2. By geographical area
3. By function.

Product

Some firms of 'general contractors' undertake a variety of work, so it is advisable to have separate groups dealing with buildings, road construction and general public works, because these activities demand very different skills and working methods. Some firms also operate complementary activities, such as a quarry, a joinery department or a plant hire section which also services other contractors. Some firms can make quite a success of such a wide range of activities and argue that, on the 'swings and roundabouts' principle, some parts of the business are always bound to be trading successfully and will even out the slack periods in the other sections. The trouble is that it is very difficult to be a 'jack of all trades', without ending up as the 'master of none', and most contractors are probably best advised to specialise in those areas of activities where they have special skills. In any event, a general contractor of this kind must have a first class accounting system which identifies his various *profit centres*, so that those which are not really paying their way can be identified before they swallow up the income generated by the more profitable parts of the business.

To aid the identification of profit centres, such a business might be divided as follows:

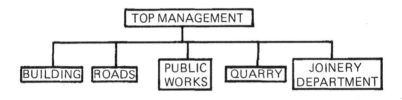

Geographical Area

Some firms, particularly those which specialise in the provision of some individual product or service, can only obtain a satisfactory level of turnover by working over a wide geographical area. Particularly in countries where transport and other forms of communication are difficult, this type of firm can only operate by setting up branches in various parts of the country and allow them quite a high degree of autonomy in managing the contracts that are obtained in their operating areas. The head office will insist on regular reporting procedures and is likely to keep a tight watch on profit levels and cash flow, but a good deal of decision-making will be delegated to the branches, so it is usually sensible to recognise this in the division of responsibilities:

257

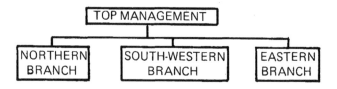

Function

For most types of contracting business, the organisation can be usefully divided by function. The simplest split is into **technical and operational tasks** on the one hand and **finance and administration** on the other as illustrated below:

Some aspects of running the business will fall quite clearly into one category or the other. For example, the updating of the books and financial records of the business is clearly a matter for the financial managers of the firm, and quality control on the site is quite clearly a technical matter. Other activities are less clear cut. For example the taking-off and ordering of materials will be done by the technical management, but negotiation of prices and credit terms as well as book-keeping and payment procedures is likely to be handled by the commercial side of the business.

Clear Division of Responsibilities

Thus, however the organisation of the firm is split up, there must be an absolutely clear division of responsibilities. This should be done by **writing down** the job description of all key staff, so that everyone knows who is responsible for what and avoid the confusion and waste of time that stems from overlapping of activities and — even more serious — delays and omission of important tasks because 'I thought someone else was dealing with it'.

Organising for Growth

Although it is not possible to prescribe an ideal organisation chart, the following pattern indicates one way of operating a growing building business. The detailed chart would have to be worked out to suit individual needs, and a small firm would combine several functions and have them run by a single manager.

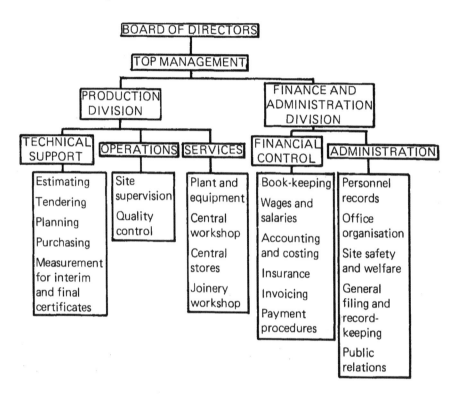

On the Site

There is also a need for a defined pattern of organisation on the site, even though every contract is a one-off job and the site staff will be dispersed to other contracts when this one comes to an end. On a small site the organisation will be very simple, perhaps consisting merely of a foreman or charge-hand who is responsible for a small group of workers. The general principle to be followed is the larger and more complex the contract, the greater is the need for specialisation in the areas of responsibility. Thus a large building or civil engineering project would be supervised by a contracts manager from head office and there would be a sizeable group of site supervisors including a site agent, sub-agents, general foreman, trade foremen and charge hands responsible for individual gangs.

The organisational problems on a large contract of this kind are as great as for the complete operations of a smaller contracting firm, and again there is a need for responsibilities to be clearly spelled out in the form of an organisation chart and detailed job descriptions. The chart for such a contract might be as follows:

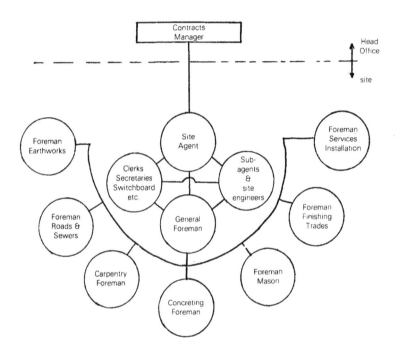

In addition to the above staff, there would probably be a quantity surveyor or measurement engineer permanently attached to the site. With such a large staff under his control, it is clear that the site agent on a contract of this size needs a considerable talent for organisation and administration as well as appropriate technical qualifications. He will also have to ensure that his own staff works closely with staff of the client's resident engineer, so that the latter are able to supervise and inspect the work in progress in accordance with the terms and conditions of the contract.

Managing to Communicate
Building is not just a technical activity. It involves working with people, both within and outside the organisation. Although it is very difficult to achieve good personal relations and the effective communication of information with a

faulty organisation structure, a good structure will not — of itself — automatically guarantee good communications. So construction managers need to understand the basic principles of good communication, if they are to become fully effective and contribute to the success of their organisation.

Formal or Informal?

There is no objection to informal communications in a small organisation. In a small firm everybody knows what is going on, and the spoken word is both quick and cheap. However, outside contacts, particularly those with the client or his representative, should always be confirmed in writing. But as the firm grows larger, the need for formal written communication grows so that the man who needs to know gets the message, and gets it right.

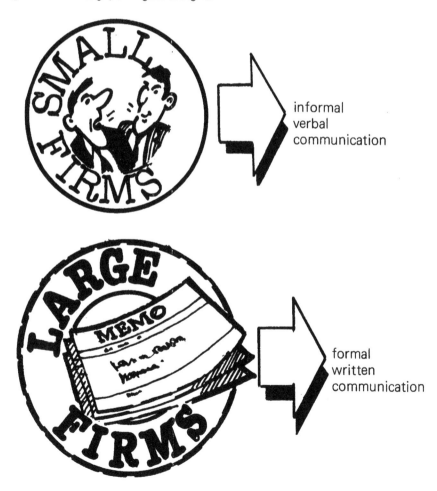

informal
verbal
communication

formal
written
communication

Is Anybody There?

For a message to get through, there must be a receiver as well as a sender. It sounds very obvious, but it is a fact that is often forgotten by inefficient communicators. To start with, both parties must be *willing* to communicate. There is much truth in the old saying that 'there are none so deaf as those who don't want to hear!'

Understanding

Communication is not an end in itself, at least as far as business communications are concerned. The object is to get the receiver of the information to respond in a positive way, and this will be demonstrated by him taking the appropriate *action*. In some cases, of course, the appropriate response will be merely to take note of the message and take physical action only if certain circumstances arise — a simple example would be a notice of how to evacuate a building in case of fire. But before the receiver can take action as desired by the sender of the message, he has to *understand* it.

Impediments to Understanding

Even if the receiver really wants to understand the message, he may still get it wrong. The receiver will attempt to interpret the message on the basis of his own background knowledge; so he may end up understanding only part of the meaning that was intended, and that may be mixed up with other meanings that were not intended. The larger the group of people receiving the information, and the longer the chain of people through whose hands the message has to pass, the greater the chances of getting things wrong. The classic story of such a misunderstanding, of a message from a general at the battle front to his headquarters is illustrated in the following picture:

The impediments may be caused by imperfections in the organisation itself, as well as by faults on the part of the sender or receiver. Good communications are achieved by limiting the receiver's attention to the relevant aspects of a particular field, and thereby eliminating all meanings that are not relevant or intended.

The Communication Process

The communication of an idea, command or fact from one person to another can be seen as a process which starts when the sender thinks of something that he wants or needs to communicate and ends (or almost ends — see below!) when the sender takes the appropriate action. The process of communication is seen most clearly in a television studio, where the 'message' is a popular programme.

Six Steps

The process starts when the author or scriptwriter thinks of an idea or basis for a documentary programme which he thinks will interest or inform the viewing public. This first stage can be regarded as the *initiation* of the message. Then the programme is played out in front of the cameras, and the message is *encoded* by them. The third stage is *transmission,* in this case over the airwaves. The fourth stage is *receiving,* and we hope that the intended receiver is tuned in to the transmitter! Then comes stage five, which takes place inside the head of the person we describe as the receiver, in which he turns the words and visual images that he perceives into an idea of his own. This stage can be described as *decoding.* The final stage is *action* and, if the programme was a documentary about the health dangers that result from smoking, successful communication will clearly have been achieved in the unlikely event of the receiver getting up from his chair and throwing a newly-bought packet of cigarettes into the dustbin!

INITIATION.   ENCODING.   TRANSMISSION. RECEIVING.   DECODING. ACTION.

The Same Pattern

The same pattern is used whenever a message is transmitted from one person to another, even by word of mouth. We start with an idea — perhaps resulting from a letter from the architect requesting that the dimensions of the rooms in a new school building should be altered to suit amended requirements by the client.

A Two-Way Process

Good communication is not just a one-way street. The sender will want to know that his message has been received and properly understood. Thus there needs to be some form of feedback to the sender from the receiver to acknowledge the message. This may take the form of the written or spoken word, or possibly the required action if it is visible to the sender. The important thing is to avoid a 'dialogue of the deaf', in which both sides talk but neither side hears!

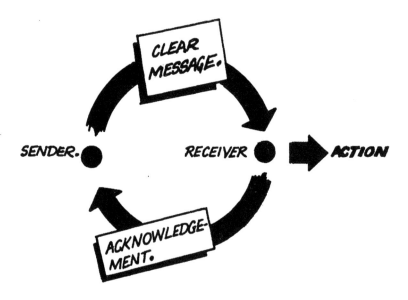

Human Aspects

Communication is not just a mechanical process. It involves human beings. Thus the good communicator takes into account the human characteristics of the receiver, and puts his points in such a way that they are both clearly understandable and acceptable. Particularly when the matter under discussion is controversial, such as the negotiation of a claim for extra costs with the client's representative, the contractor will need to take extra care to put his argument in a way that is likely to make the receiver sympathetic.

Agreement First

One good way of making the man on the other side of the table receptive to your ideas is to put the good news first, by starting your letter or verbal presentation with points of agreement before going on to discuss other matters. In this way, the receiver starts off by nodding his head and is at least in the right mood when he comes to the rest of the message.

Negotiation

Where the communication takes the form of a formal negotiation, e.g. with the client over a negotiated addition to the contract or a supplier over discount and credit arrangements, there is a real need to put yourself in the other man's shoes and be sensitive enough to pick up an idea of just how far he will be prepared to go. Brilliant negotiators are born rather than made, but with practice most managers can put up at least an adequate performance. The secret weapon can be thorough research and planning, so that you walk into the conference room knowing as much as possible about the subject under discussion, the other man's attitude and business background and have a 'game plan' to guide you in putting your case in a logical and persuasive way. As with learning to ride a bicycle, practice makes perfect. The experienced negotiator will be so much in charge of his own emotions that he will have a 'feel' for the right moment to be charming, to get angry or even to jump up from his chair and start to walk out of the room. The tricks of the trade in negotiating a simple purchase from the stallholder in the local market can also be of service in much more important business negotiations!

Listen!

The author's favourite definition of the word 'tact' is 'the ability to see things from the other man's point of view'. If you are going to hope to understand his viewpoint, you must first find out what he has to say. So the first commandment for the negotiator must be — *listen* and learn. In the words of the old proverb, 'we were given two ears and only one mouth, because God believes it is more important to listen than to talk!'

Channels of Communication

Information can be transmitted in a number of ways, and each has its own advantages and disadvantages. The first choice is between written and verbal contact. Letters and memos have the advantage that the story can be put clearly in black and white, and a copy can be made for record purposes, but they are of course slower to get a response (much slower if the postal service is bad!) and less personal than a visit or a telephone call (assuming you can get through to the person you are calling!).

Letters

Letters are a vital medium of communication where the message has to be clearly understood, and where a record is required. For the receiver, they have the advantage that they can be read at his convenience, and he too has a written statement of what the sender wants, to which he can refer later if required. For a contractor, the letters he writes to his client also say something about him as a businessman. So it is worth setting them out neatly on headed notepaper, as scruffy correspondence is often taken as a sign of a slip-shod approach to other aspects of business practice. When the letter has been drafted, read it through *as if you had just received it* through the post. Does it make sense? Does it assume some knowledge that the sender has but the receiver does not? Is it clear and unambiguous? Remember the four qualities that mark a good business letter. It should be:

— clear
— short
— concise
— precise.

Letters that concern quotations and contracts, or may have any kind of contractual or legal implication, either with clients or suppliers, need special care. Read them. Then re-read them. And then imagine, if the worst should happen, how they would look if they were to be read by an arbitrator or a judge.

Memoranda

For less formal written communications, particularly within the organisation, memoranda can be used. They are also useful on the site for simple exchanges of information, such as a request for additional details from the clerk-of-works. The copy memo may be invaluable later in helping to establish a claim for extra payment or an extension of the contract period.

Drawings, Schedules and Contract Documents

These items are usually provided by the architect or engineer, and the contractor has the task of interpreting them correctly and getting the information to those of his staff or his sub-contractors who will be doing the work on the site. A strict registration procedure should always be followed and the new drawing or document should be date stamped as it is received. All but one of the superseded copies of amended drawings should be retrieved from the site, and that one copy should be clearly marked 'SUPERSEDED' and filed well away from current drawings. A contractor can lose a great deal of money through the wasted effort of his staff working to drawings that have been superseded or withdrawn. When an amended drawing is received, check if the alteration will require any demolition or extra costs through redundant and unsaleable materials. If this is the case, get the clerk-of-works to give a written confirmation of this *immediately,* as it will be impossible to substantiate the facts later in the contract.

Telegrams and Cables

If the message is really vital, it can sometimes pay to spend the extra and put it in the form of a telegram. Besides being quicker, a telegram has an electrifying effect on the most lethargic of correspondents (particularly if sent 'reply-paid').

Personal Visits

A face-to-face meeting takes more time, but it is more flexible and allows the participants to 'test out the ground' while the matter is explained and discussed. If properly handled, they also improve interpersonal relationships as one can assess personalities and attitudes. Important issues can and should be confirmed in writing afterwards.

Meetings

A personal visit is in effect a meeting between two people. But most people think of meetings as more formal occasions which allow a group of people to be informed, to exchange information and reach decisions simultaneously. Most contractors are at least involved in site meetings, which are usually called on a monthly basis to discuss progress and any outstanding difficulties and problems. These meetings are usually seen as a chore, and sometimes as a minefield of potential embarrassments when work is behind schedule and everyone is looking for someone to blame.

Opportunity

But the site meeting should also be seen as an opportunity; both to obtain the outstanding information that the contractor needs and to secure the assistance of the client's representative in dealing with hold-ups caused by third parties. They also offer the contractor the chance to portray himself in a favourable light, by highlighting his successes and the quality of his plans for future progress. Even if the client is not personally represented at the meeting, his representative will report back. So at every site meeting there is a marketing opportunity waiting to be grasped.

Reports

The contractor will usually be expected to make a report on progress, and on larger contracts the report should be presented in writing. Whether written or verbal, the report should be clear and logical and no longer than is necessary to convey all the information that is required. It is a mistake to assume that people are more impressed with long reports, so irrelevant information and other 'padding' should be ruthlessly cut out. Every report should answer four questions by being:

To	someone	*Who?*
From	someone	*Who?*
About	something	*What?*
At	some time	*When?*

Purpose

If a meeting is to be effective, its purpose must be clearly stated far enough in advance for all the participants to come fully prepared. If written reports and papers are to be considered at the meeting, they should be circulated to the participants in advance. There should always be a written agenda, and a standard format is useful for regular meetings such as monthly site meetings.

Who Should Attend?

'The more the merrier' is a good slogan for a party, but the worst possible principle for a business meeting. Attendance should be strictly limited to those who have a contribution to make or who really need to be present at the discussion, as distinct from reading the decisions when the minutes of the meeting are circulated. At a site meeting, the only sub-contractors who need to attend are those with whom there are difficulties; the rest would be better employed elsewhere.

Running the Meeting

The architect usually acts as chairman at site meetings, but the contractor may well run meetings with his sub-contractors, suppliers, his own staff or perhaps in his local contractors' association. A good chairman ensures that everyone sticks to the items as set out on the agenda, and sets a good example by never wandering from the point himself. If there is an item for 'contract progress', any points relating to that should be ruled out of order until that item is reached. Most people prefer a firm chairman, particularly if the firmness is tempered by a pleasant and tactful approach. One item that often gives trouble is 'matters arising'. This must be strictly interpreted as *matters arising from the minutes of the last meeting.* It does *not* mean any odd things that comes into somebody's mind. Other matters, even if they are worthy of discussion, should be kept for main agenda items or 'any other business'. This latter item must be regarded as 'any other relevant business', and should not be used by participants to spring surprises on the others by avoiding the pre-publicity of reserving a full agenda item. Matters of this kind should be left over to the next meeting, or the meeting can be adjourned to give people a chance to consider all the ramifications of the proposed course of action.

Where do we go from here?

Meetings, like other forms of communication, are point-less if they do not result in *action*. At the end of the meeting, there must be no doubt on *what* has been decided, *what* form the action is to take and *who* will be responsible for implementation (and usually *when* is the deadline for doing it). Thus it is always necessary to confirm what was agreed by issuing 'minutes' as soon as possible after its conclusion. Agree in advance who is to act as secretary and prepare the minutes. As a protection, every party present would be wise to keep their own notes in case of later disagreement with the 'official' minutes.

Contents of Minutes

Minutes should not read like a novel, with a verbatim account of what everybody had to say. They should, however, record in a concise way all the *decisions* taken by the meeting and, where appropriate, the facts or evidence that led to the decision. They will be subject to ratification as 'a true and correct record' at the following meeting, but should nevertheless be issued and acted on as soon as possible after the meeting as speedy decisions and *action* is vital if the contract is not to be delayed. To ensure that there is no confusion about who is to do what, an 'action' column can be included and the initials of the person who is responsible for implementing each decision should be noted there.

MINUTES OF A MEETING HELD AT THE SITE OF XXXXXX HEALTH CENTRE ON 28.11.79 AT 2.30 pm

Present: xxxxxxx (Chairman)
xxxxxxx
xxxxxxx
xxxx

DECISION	ACTION
1. xxxx xxx xxxxx xx xxxxx xxx xx xxxx xx xxxx xxxxxxx xxxxxxx xxx xxx xxx	AB
2. xxxxx xxxxx x xxxx xxxx xxx	CD
3. xxxx xxx xxxxx xx xx xxxxxx xxxx xx xxx xxxxx xxxx xx xxxxx xxxxx xx x	EF/GH

www.ingramcontent.com/pod-product-compliance
Lightning Source LLC
Jackson TN
JSHW011409130125
77033JS00024B/939